Scientific Research Methodology

This book serves as a comprehensive overview of the principles, tools, and techniques of scientific research methodology. It guides readers through various research methodologies and techniques, starting from the basics and culminating in scientific writing and publication. Using simple language, this well-rounded resource empowers readers of all levels – from beginners to seasoned experts – to publish their research outcomes effectively. Some of the key features are as follows:

- Offers a thorough exploration of research methodologies, literature survey strategies, and classification of research approaches
- Explains fabrication and manufacturing techniques and advanced characterization methods to bolster key concepts and practical skills
- Emphasizes data optimization and post-processing techniques and introduces state-of-the-art tools for effective analysis and interpretation of data, including a brief introduction to MATLAB, Python, machine learning, and artificial intelligence
- Discusses documentation and scientific writing, offering invaluable advice for creating and publishing impactful scientific manuscripts

Covering all essential steps necessary for conducting technical research, this book is particularly tailored to benefit new researchers in science and engineering fields engaged in academic studies and technical writing.

Scientific Research
Methodology
Principles, Tools, and Techniques

Sanjeev Singh Yadav, Shivani Gupta, and
Atul Kumar Choudhary

CRC Press
Taylor & Francis Group
Boca Raton London New York

CRC Press is an imprint of the
Taylor & Francis Group, an **informa** business

Designed cover image: © 2026 Shutterstock

First edition published 2026
by CRC Press
2385 NW Executive Center Drive, Suite 320, Boca Raton FL 33431

and by CRC Press
4 Park Square, Milton Park, Abingdon, Oxon, OX14 4RN

CRC Press is an imprint of Taylor & Francis Group, LLC

© 2026 Sanjeev Singh Yadav, Shivani Gupta, and Atul Kumar Choudhary

ISBN: 978-1-041-12458-0 (hbk)
ISBN: 978-1-041-12457-3 (pbk)
ISBN: 978-1-003-66487-1 (ebk)

DOI: 10.1201/9781003664871

Typeset in Palatino
by SPi Technologies India Pvt Ltd (Straive)

This book is dedicated to all the researchers/scientists.

Contents

Foreword

Down through the ages, scientific curiosity has proved to be a powerful tool, paving the way for unraveling the mysteries of Nature and thereby helping enhance the quality of human life. Titled *Scientific Research Methodology: Principles, Tools, and Techniques*, this book will serve as a comprehensive guide, bringing clarity and showing the path to the often arduous journey of scientific investigation. In this digital era, data-driven decisions and evidence-based policies dominate public discourse. Consequently, the ability to critically analyze and evaluate information and design is construed as a vital requirement more than ever. This book will impart the reader fundamental knowledge and tools that are required to navigate this journey with confidence. This book is a comprehensive mix of theoretical principles and practical knowledge, making it an invaluable tool to students and young researchers seeking a deeper understanding of scientific research methodologies.

The authors have made a sincere effort in presenting the complex subject with clarity and in an easy-to-understand language without robbing it of its significance. They have dealt with the subject not as a set of procedures but as a discipline that requires curiosity, patience, and precision. As you turn these pages, not only will you gain an insight into the fundamentals of scientific research, but you will also have an appreciation for the intellectual rigor and ethical responsibility that are needed to make meaningful discoveries. It is a book that will serve both as a toolkit and as a compass for anyone embarking on the journey of scientific exploration. I hope it will reach a large audience and fulfill its intended purpose.

Dr. V. Sampath
Former Professor
Department of Metallurgical and Materials Engineering
Indian Institute of Technology Madras, Chennai

Preface

Identifying correct research approaches and suitable characterization techniques has always played a critical role in human progress since prehistoric times in terms of the development of new technologies. Research methodology and techniques have gained significance as an essential branch, since they are directly related to different engineering branches and other fields like medicine, forensic science, and economics. The existence of the advanced technology that we see today has all been possible mainly due to the advances in research in various sectors like agriculture, transportation, energy, telecommunication, construction, aerospace, computers, education, and many others that have significantly impacted the lives of everyone in society. The understanding of several methodologies/techniques is vital for conducting research from scratch to the scientific writing/publication process. Various steps are required to understand before conducting any technical research. Also, in-depth knowledge of research methodologies, literature survey techniques, and advanced characterization methods with a solid foundation and understanding of essential concepts and practices is needed for carrying out the research. Researchers should know the various fabrication and manufacturing techniques, as well as material response assessment methods that can be directly applied to real-world research and development projects. The various optimization and post-processing techniques and the introduction to cutting-edge tools and methods enhance their ability to optimize and analyze data effectively.

This book covers all the aspects mentioned above to prepare a strong research foundation that is important for advanced material design and development. The main purpose of the book is to provide a holistic picture and to give a clear understanding of research methodology and techniques in a simple language at a competitive cost. Chapter 1 highlights the significance of research methodology and its introduction. Chapter 2 deals with the different approaches to literature surveys. Some commonly used material assessment/evaluation techniques are explained in Chapter 3, and applications of artificial intelligence (AI) and machine learning (ML) for research have been discussed in Chapter 4. The various manufacturing/fabrication processes are presented and discussed in Chapter 5, highlighting the significance and role of process parameters for optimizing the fabrication processes. Several characterization techniques are discussed in Chapter 6. Chapter 7 discusses the introduction to advanced tools/programming languages such as MATLAB, Python, and various optimization techniques in a simple way with proper schematic diagrams. Finally, Chapter 8 introduces the world of scientific writing to the readers/researchers.

Although some books are available on the topic, such as *Scientific Research Methodology*, in the market, the proposed book highlights the subject's significance by correlating it with practical applications that make readers realize the importance of the subject. Some salient features of this book are as follows:

- The significance of the subject and its contents is highlighted by correlating with practical applications that make readers realize the importance of the subject.
- The explanation of the contents and the presentation of the subject are easy to understand.
- All concepts and explanations are supported with relevant schematic diagrams and well-designed figures for better clarity and easy understanding.
- Unit-wise conceptual, theoretical questions and practical numerical problems (short and comprehensive) are given for students' self-evaluation.
- We hope this book serves as a comprehensive guide for students, scientists, and researchers on their journey toward rigorous and impactful scientific inquiry.

Sanjeev Singh Yadav
Shivani Gupta
Atul Kumar Choudhary

Authors

Sanjeev Singh Yadav, PhD, is a researcher in the field of mechanical engineering with a focus on the fatigue behavior of metallic materials. He earned a PhD at the Indian Institute of Technology Ropar, India. Prior to pursuing his doctoral studies, he served as an assistant professor for three years at various engineering institutions. He earned an MTech in computer integrated manufacturing from National Institute of Technology Warangal, Telangana, and a BE in mechanical engineering from Rajiv Gandhi Technical University, Bhopal, Madhya Pradesh. Dr. Yadav is the author of *Materials Science and Engineering*, published by Cambridge Scholars Publishing, Newcastle, United Kingdom. His research contributions include several publications in reputed international peer-reviewed journals, primarily regarding the fatigue behavior of metallic materials.

Shivani Gupta, PhD, earned a PhD (2024) in production engineering at the Indian Institute of Technology, Roorkee, India, and an ME in mechanical engineering at the National Institute of Technical Teachers Training and Research, Chandigarh. She has two years of teaching experience as a lecturer at the Bundelkhand Institute of Engineering and Technology, Jhansi, Uttar Pradesh, from 2012 to 2014. She holds a Gold Medal (Honor) in her master's degree, Panjab University, Chandigarh, in 2016. She also has one year of research experience at the renowned CSIR–Central Scientific Instruments Organisation Chandigarh. Dr. Gupta has international research exposure at the Materials Research Institute, Pennsylvania State University, USA, as a visiting research scholar from August 2021 to April 2022. Dr. Gupta is presently an assistant professor in the Mechanical Engineering Department at Dr. Vishwanath Karad MIT World Peace University, Pune, Maharashtra, India. Her research interests include advanced manufacturing processes, nano-biocomposites, biocompatible healthcare devices, microwave materials processing, additive manufacturing, material characterization, modeling, and simulation. She has published six SCI journal articles, five Scopus research articles, and two book chapters in reputed journals.

Atul Kumar Choudhary, PhD, earned a PhD in mechanical engineering at the Indian Institute of Technology, Bhilai (2018–2023), and an MTech in material technology at the National Institute of Technology Warangal (2013–2015). He has two years of teaching experience as an assistant professor at Mumbai University (2016–2018). He earned a BE in mechanical engineering (Honors) at Chhattisgarh Swami Vivekanand Technical University, Bhilai, in 2013. He worked at the renowned CSIR–National Metallurgical Laboratory Jamshedpur for a year (2014–2015) in the MST Division under Dr. Sivaprasad. His research interests include modeling and simulation in friction stir welding, friction stir processing, additive manufacturing, and materials characterization. He has published six articles in SCI journals and two book chapters.

Acknowledgments

First and foremost, we would like to express our heartfelt gratitude to the Almighty of this world and our family members for their patience, moral support, and cooperation during the writing of this book. This project would not have been possible without their support and encouragement; we would always be indebted to them. We have been fortunate to have, and continue to have, great teachers, colleagues, and friends. We want to thank all the teachers, friends, and all our well-wishers who have been great support throughout the period. This book would not be possible without discussions with our professional colleagues. We want to express our heartfelt gratitude to **Dr. Jose Immanuel R.**, Indian Institute of Technology, Bhilai, for valuable suggestions, and the following reviewers for reviewing the book's chapters.

Dr. Dhruva Kumar Goyal (Chapter 1), Assistant Professor at Amity University, Mohali, Punjab, India

Dr. Vemuri SRS Praveen Kumar (Chapter 2), Head of New Product Development, Asahi India Glass Ltd., Navi Mumbai, Maharashtra, India

Mr. Rajat Dhiman (Chapter 3), Senior Research Fellow, Indian Institute of Technology Ropar, Punjab, India

Dr. Sunita Parinam (Chapter 4), Assistant Professor, MIT ADT University, Pune, Maharashtra, India

Mr. Uday Kumar Paliwal (Chapter 5), Former Assistant Professor, L.T. College of Engineering, Navi Mumbai, Maharashtra, India

Professor V. Sampath (Chapter 6), Former Professor, Indian Institute of Technology Madras, Tamil Nadu, India

Mr. Sumit Devraye (Chapter 7), Assistant Manager, Accenture Private Limited, Indore, Madhya Pradesh, India

Mr. Chirag Panwariya (Chapter 8), Senior Research Fellow, Indian Institute of Technology Roorkee, Uttarakhand, India

We extend our heartfelt gratitude to **Ms. Allison Shatkin** (Senior Publisher at Taylor & Francis) for accepting the proposal, **Ms. Ariel Finkle** (Editorial Assistant at Taylor & Francis) and **Ms. Lekshmi Priya** (Project Manager at Straive) for their help throughout the publishing process. The suggestions and comments on the manuscript by *anonymous reviewers* are also gratefully appreciated. Lastly, we would like to thank everyone who has, directly and indirectly, contributed to this book.

1

Research Methodology

1.1 Introduction

Research methodology is the structured process of investigating a specific topic or question to uncover insights, test hypotheses, or develop theories. It encompasses the principles, strategies, and techniques employed to design and execute research effectively. The methodology serves as the blueprint for the research process, guiding researchers in systematically collecting, analyzing, and interpreting data to ensure that the findings are reliable and valid. The research objectives, the nature of the problem, and the type of data required influence the choice of methodology. Broadly, research methodologies are categorized into qualitative, quantitative, or mixed methods explained below and as shown in Figure 1.1. This is an important tool for every researcher and scientist to follow a particular guideline/policy before carrying out the research work. Additionally, research methodology minimizes errors and biases, ensures that ethical guidelines are followed, and helps maintain scientific rigor. It provides a roadmap for the entire research process, from identifying the research problem to collecting and analyzing data, and ultimately concluding. This structured approach not only saves time and resources but also increases the reliability and validity of the research outcomes.

1.1.1 Quantitative Research

This focuses on numerical data and employs statistical tools to identify patterns, test hypotheses, and make predictions. It is often used in scientific and empirical studies where measurement and objectivity are critical. Quantitative research is a systematic investigation that focuses on gathering numerical data and analyzing it to uncover patterns, relationships, or trends. This type of research is grounded in the scientific method and relies

DOI: 10.1201/9781003664871-1

Research Methodology

1. Start Identify Research Problem

↓

2. Choose Research Approach

Quantitative (Numerical, Structured)

- Data Collection: Surveys, Experiments
- Data Analysis: Statistical Tools (Excel or Python)
- Outcome: Numerical Findings, Hypothesis Testing

Qualitative(Descriptive, Exploratory)

- Data Collection: Interviews, Focus Groups, Case Studies
- Data Analysis: Thematic, Content Analysis
- Outcome: Insights, Patterns, Subjective Interpretation

Mixed Methods (Combination of Both)

- Data Collection: Both Surveys & Interviews
- Data Analysis: Integrating Statistics & Themes
- Outcome: Comprehensive Understanding

↓

3. Interpretation & Conclusion

↓

4. End: Research Findings & Implications

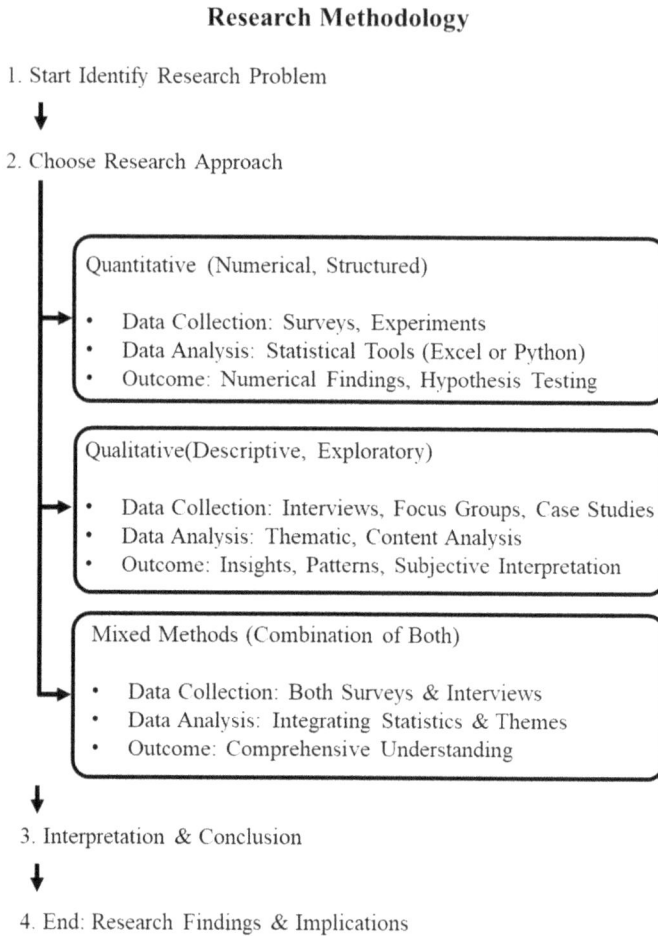

FIGURE 1.1
Flowchart representing the steps in research methodology.

on statistical tools to test hypotheses and generalize about a population. Researchers in quantitative studies collect data using instruments like surveys, experiments, or observations, ensuring that the data can be measured and expressed in numerical terms. The goal is to achieve objectivity and reliability, often by using large sample sizes to ensure that findings are statistically significant. The results of quantitative research are typically presented through charts, graphs, and statistical analyses, offering a clear and concise picture of the research outcomes.

1.1.2 Qualitative Research

It explores complex phenomena through non-numerical data like interviews, observations, and texts. It aims to understand concepts, perceptions, and experiences, providing rich and in-depth insights. Qualitative research is a method of investigation that focuses on understanding human practices, behaviors, and social phenomena from a subjective and in-depth perspective. Unlike quantitative research, which seeks to quantify data, qualitative research explores the meanings, interpretations, and contexts behind individuals' actions or beliefs. This type of research often involves methods like interviews, focus groups, case studies, or participant observations, where researchers engage directly with participants to gather rich, descriptive data. The aim is to capture the complexity of a situation, often in its natural setting, and to provide insights into how and why things occur. The findings are typically analyzed thematically, and the results are presented in narrative form, emphasizing understanding over-generalization. Qualitative research is particularly valuable in exploring new areas of study where little is known or in situations where human experiences and perspectives are central to the research.

1.1.3 Mixed Methods

It integrates both qualitative and quantitative approaches to leverage the strengths of each, offering a comprehensive understanding of the research problem. A mixed-methods approach in research combines both quantitative and qualitative methods to provide a more comprehensive understanding of a research problem. By integrating numerical data with rich, descriptive insights, researchers can draw on the strengths of both approaches. The quantitative component typically involves collecting numerical data through surveys, experiments, or statistical analysis, which allows for generalizability and objective measurement. On the other hand, the qualitative component involves gathering non-numerical data, such as interviews, observations, or open-ended responses, which offer deeper insights into participants' experiences, beliefs, and motivations. This combined approach enables researchers to triangulate findings, validate results, and explore the research topic from multiple angles, leading to more robust and nuanced conclusions. Mixed-methods research is particularly valuable when addressing complex questions that require both broad patterns and in-depth understanding. A well-defined research methodology ensures that the study is methodologically sound, ethically conducted, and yields results that contribute meaningfully to the field of study. It also enhances the replicability of the study, allowing other researchers to validate or build upon the findings.

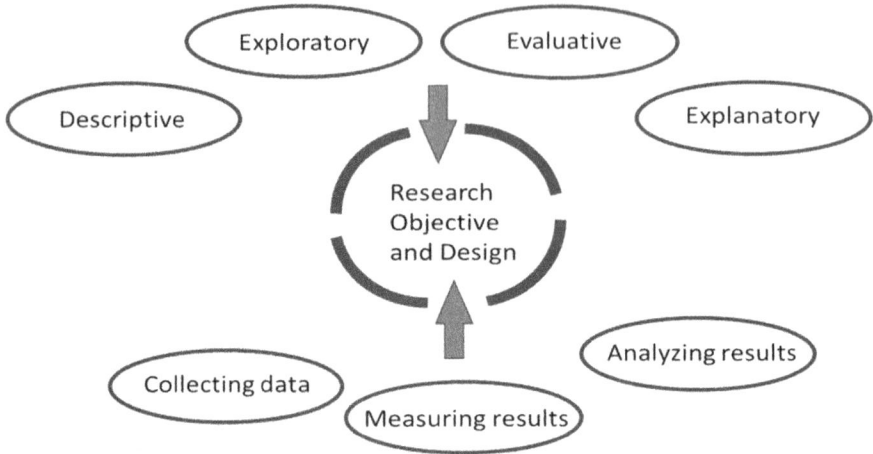

FIGURE 1.2
Flowchart representing the steps of the research objective and design.

1.2 Research Objective and Design

1.2.1 Research Objective

The research objective defines what the researcher aims to accomplish through the research study. It is a concise and clear statement that plans the specific objectives or purposes of the research. Research objectives provide direction and focus to the study, helping to determine the research questions, hypotheses, and overall approach. A flow diagram is shown in Figure 1.2. The research objectives should be specific and clearly define what the researcher intends to investigate. It should be measurable so the researcher can assess earlier whether the goals can be achieved with justifiable assumptions. The objectives should align with the larger goals of the research and be significant to the field of study. Also, it should be achievable and realistic within the limitations of resources, time, and the researcher's expertise. Types of research objectives are described as follows:

1. **Descriptive** – Aimed at describing the characteristics of a phenomenon or population. It aims to provide an accurate and detailed account of a phenomenon, situation, or population. The goal is to describe the characteristics or features of a subject without manipulating the environment or variables. It helps researchers to answer questions such as "What?" and "How?" For example, a descriptive study might look at the demographics of a population or the trends in consumer behavior over a period of time.

2. **Exploratory** – Focused on exploring new areas where little information is available. They are used when little is known about a topic,

and the researcher seeks to explore it more deeply. The goal is to gain insights and understand the underlying factors or issues. This type of research is often conducted in the early stages of a study to identify potential variables or hypotheses. It is flexible and open-ended, helping to guide future, more focused research. An exploratory study might involve gathering qualitative data through interviews or case studies to uncover new ideas or patterns.

3. **Explanatory** – Aimed at explaining the causes or reasons behind a particular phenomenon. Aim to explain the causes and effects of a phenomenon. This type of research is concerned with identifying relationships between variables and understanding how and why something happens. Explanatory research typically builds upon exploratory findings and seeks to test hypotheses or theories. It addresses questions such as "Why does this occur?" and "What are the reasons behind the patterns observed?" Experimental studies are often used for explanatory research to establish causal relationships.

4. **Evaluative** – Concerned with assessing the outcomes or impacts of a particular intervention or program. This focuses on assessing the effectiveness, impact, or quality of a program, policy, product, or intervention. The goal is to evaluate whether the objectives of a project or initiative have been achieved and to measure its overall success. Evaluative research answers questions such as "How well does this work?" and "What are the outcomes?" This type of research is crucial in areas like education, healthcare, and business, where assessing the value and outcomes of specific interventions or programs is essential for decision-making.

1.2.2 Research Design

The research design is the universal approach or blueprint that the researcher uses to incorporate the distinctive factors of the findings logically and rationally. It outlines the procedures for collecting, measuring, and analyzing results.

1. **Collecting data** – It involves the systematic gathering of information from various sources, such as surveys, interviews, observations, or experiments. The collection process is carefully planned to ensure that the data represents the research question and the population being studied. This step must be methodical to avoid bias and ensure consistency across all data points.

2. **Measuring results** – It refers to the process of quantifying or categorizing the data collected in a way that allows for comparison and analysis. Measurement tools, such as scales, instruments, or questionnaires, are employed to ensure that the data is accurately captured and aligned with the research objectives. The choice of

measurement techniques depends on the type of research, whether quantitative or qualitative.

3. **Analyzing results** – It involves examining the collected data to identify patterns, relationships, or trends. Statistical methods or qualitative coding techniques are used to process the data and interpret the findings. Analysis helps to draw conclusions, test hypotheses, and provide insights that answer the research questions. This step is essential for translating raw data into meaningful information that contributes to the research field.

1.2.3 Relationship Between Research Objective and Research Design

The research objective determines the type of research proposed that is suitable for the investigation. For instance, if the purpose is to explore the correlation between two quantities, a correlational model might be appropriate. If the objective is to determine the impact of an intervention, an experimental design would be more suitable. The design chosen must align with the objectives to ensure that the research effectively addresses the questions posed.

1.3 Methods of Data Collection

Data collection is a prominent step in the research procedure, as it includes collecting the evidence and records needed to answer the research queries, test assumptions, and accomplish the research aims. Different methods of data gathering are used depending on the type of investigation, the research strategy, and the nature of the data required. Here's an overview of common data collection methods, as shown in Figure 1.3.

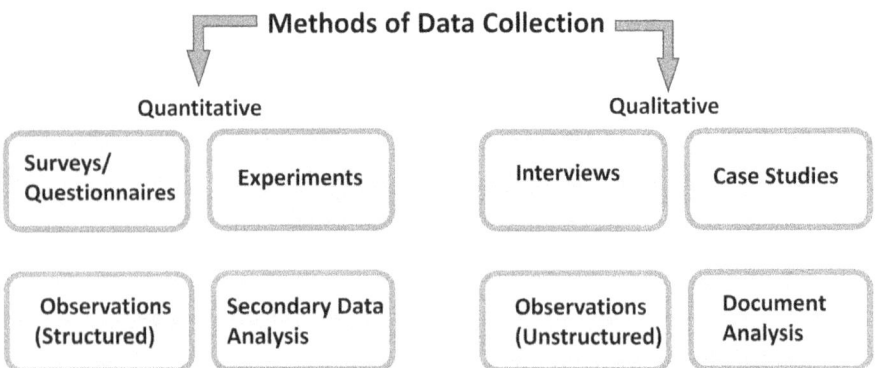

FIGURE 1.3
Flowchart representing methods of data collection.

1.3.1 Quantitative Data Collection Methods

Quantitative methods are used to collect numerical records, which can be measured and analyzed statistically.

1. **Surveys/Questionnaires** – Here, structured instruments with a series of questions are designed to gather quantifiable information during research. These can be used in descriptive, correlational, and experimental research to collect data from a large sample. This method is efficient for large samples, easy to analyze, and can be distributed online or on paper, but it has the disadvantage of limited depth of response, the potential for low response rates, and possible bias in self-reported data.

2. **Experiments** – Controlled studies are systematic experiments in which researchers deliberately alter one or more independent variables to examine their impact on one or more dependent variables. These studies are carefully designed to isolate and understand causal relationships by minimizing external influences. By manipulating specific factors while keeping other conditions constant, researchers can observe how changes in the manipulated variables directly affect outcomes, providing valuable insights into the dynamics of the phenomenon studied. They are used in experimental research to establish causal relationships between variables. These have added advantages of high control over variables and strong evidence of causality. However, they may lack ecological validity, can be resource-intensive, and ethical considerations may arise.

3. **Observations (Structured)** – Systematic observation and recording of behaviors or events using predefined categories. It is used in studies where specific behaviors or events need to be quantified. It has added advantages that provide objectified data, which is useful for studying natural behaviors. It has limitations as the observer bias is limited to observable behaviors and may not capture context.

4. **Secondary data analysis** – Here, the analysis of existing data collected by other researchers or organizations is generally used when primary data collection is not feasible or when additional analysis of existing data is needed. It has added advantages of cost-effectiveness and time-saving and allows the analysis of large datasets. However, this method has limited control over data quality and may not perfectly match the research objectives.

1.3.2 Qualitative Data Collection Methods

Qualitative techniques are used to gather non-numerical information, providing deeper insight into people's experiences, attitudes, and behaviors.

1. **Interviews** – Interviews are a qualitative research method concerning direct, one-to-one, or group discussions between the researcher and participants. They can be categorized into three main types based on their structure. Structured interviews adhere to a predetermined set of questions, ensuring consistency across participants. Semi-structured interviews follow a general guide but allow flexibility for the researcher to probe deeper into specific topics as needed. In contrast, unstructured interviews are open-ended, with no predefined questions, allowing for a free-flowing conversation that can explore unexpected insights. Interviews are widely used to gain an in-depth understanding of participants' perspectives, making them a valuable tool for uncovering detailed and nuanced information.

2. **Observations (Unstructured)** – Observation consists of observing and recording actions or events in their natural settings without relying on predefined categories. This method is commonly employed in ethnographic or exploratory research to gain authentic insights into natural behaviors as they occur. One of its primary advantages is the ability to capture data in a real-world context, offering a rich and in-depth understanding of the phenomenon under study. However, observation also has its challenges, including the potential for observer bias, the time-intensive nature of the process, and the lack of a predefined structure, which can make data analysis more complex.

3. **Case studies** – A case study involves an in-depth analysis of a single circumstance or a lesser number of cases within their real-world context. This method is frequently utilized to explore complex issues in detail, particularly in fields such as social sciences, business, and psychology. One of its key advantages is the ability to provide a comprehensive understanding of the subject while situating findings within their specific context. However, case studies have limitations, including limited generalizability due to the focus on a small number of cases and the potential for researcher bias, which can influence interpretation.

4. **Document analysis** – Document analysis is a systematic method of reviewing and interpreting existing materials such as reports, letters, diaries, and other written records. It is commonly used in historical research, policy analysis, and as a supplementary tool alongside other qualitative methods. This approach offers several advantages, including the ability to provide historical or contextual insights and its non-intrusive nature, as it relies on pre-existing documents. However, its limitations include the potential scarcity or unreliability of documents and the possibility of interpretive bias, which can affect the analysis and conclusions.

1.3.3 Mixed-Methods Data Collection

A mixed-methods study integrates both quantitative and qualitative data collection approaches, leveraging the powers of each to provide a more complete understanding of the research problem. This methodology can follow two primary designs: sequential designs, where data is collected in phases (quantitative first, followed by qualitative, or vice versa), and concurrent designs, where both types of data are collected simultaneously. Mixed-methods research offers significant advantages, such as delivering a more complete picture of the phenomenon under study and enabling cross-validation of findings. However, it also poses challenges, including the complexity of design and analysis and the need for expertise in both qualitative and quantitative methods.

Choosing the right data collection method depends on several factors. These include research objectives, which define what the study aims to uncover; the research design, determining whether a qualitative, quantitative, or mixed-methods approach is most suitable; the population being studied, considering participant characteristics and accessibility; the resources available, such as time, budget, and tools; and ethical considerations, ensuring participant rights and protections. Each method has unique strengths and limitations, and the choice should align with the research questions and overall study design to ensure effective and ethical data collection.

1.4 Processing and Analysis of Data

The processing and analysis of data are serious stages in the research process, where the raw data collected through various methods is systematically examined to get a meaningful conclusion with a good description. Here's an overview of these stages as shown in Figure 1.4.

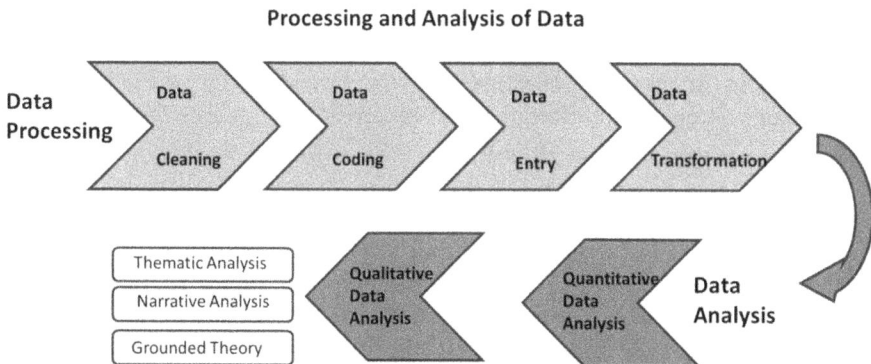

FIGURE 1.4
Flow process representing the processing and analysis of data.

1.4.1 Data Processing Techniques

Data processing involves preparing the collected data for analysis. This step is important to confirm that the data is complete, accurate, and in a proper format for investigation.

1. **Data cleaning** – The procedure of detecting and removing errors or variations in the data is crucial to improving the quality of research outcomes. This process, often referred to as data cleaning, involves several tasks aimed at ensuring the accuracy and consistency of the data before analysis. One key task is handling missing data, where researchers must identify any missing values and decide on an appropriate strategy – whether to remove, impute, or ignore them based on their impact on the analysis. Another important task is removing outliers, which involves detecting extreme values that could distort the results and determining whether they should be excluded or adjusted. Correcting errors is another critical step, where researchers fix data entry mistakes such as incorrect values or duplicate entries, ensuring the integrity of the dataset. Lastly, standardizing data ensures that the information is in a consistent format, such as converting dates into a uniform style or ensuring measurement units are aligned, which makes it easier to compare and analyze. By addressing these tasks, researchers can improve the quality of their data, leading to more accurate and reliable research outcomes.

2. **Data coding** – Data coding is an essential step in data analysis that involves transforming qualitative information into a structured format that can be easily examined and interpreted. This process typically includes classifying responses, appointing numerical or symbolic codes, and organizing data in a way that facilitates statistical or thematic analysis. By converting open-ended responses, interviews, or textual data into quantifiable units, researchers can identify patterns, trends, and relationships within the dataset. Effective coding ensures consistency and accuracy, allowing for meaningful comparisons and deeper insights. It is widely used in fields such as social sciences, marketing research, and machine learning (ML), where large volumes of qualitative data need to be systematically analyzed.

3. **Data entry** – Data entry is the process of inputting collected information into a computer system or software for further analysis and processing. It serves as a crucial step in data management, ensuring that raw data is accurately recorded and organized for future use. This process can be carried out manually or through automation, depending on the volume and complexity of the data. Manual data entry involves individuals inputting information from paper-based sources, such as surveys, forms, or handwritten notes, into a digital

database. While this method allows for careful verification, it can be time-consuming and prone to human error. On the other hand, automated data entry utilizes software tools, such as optical character recognition (OCR) or data capture systems, to efficiently extract and store information with minimal human intervention. Automation improves speed, reduces errors, and enhances overall efficiency. Regardless of the method used, accurate data entry is essential for reliable data analysis, decision-making, and business operations across various industries.

4. **Data transformation** – Data transformation is a critical process in data preparation that involves modifying raw data to ensure that it aligns with the requirements of analysis. This process enhances the quality, consistency, and usability of data, making it more suitable for deriving meaningful insights. One key aspect of data transformation is aggregating data, where detailed records are summarized at a higher level, such as converting daily sales figures into monthly totals to identify trends over time. Another important technique is normalizing data, which involves scaling values to a common range or format, making comparisons more reliable, especially when data comes from different sources. Additionally, creating new variables from existing data is often necessary to derive more insightful metrics, such as computing a profitability ratio from revenue and costs or developing an index that combines multiple indicators. Effective data transformation ensures that datasets are structured optimally, improving the accuracy and efficiency of statistical modeling, ML, and business intelligence applications.

1.4.2 Data Analysis Techniques

Data analysis involves thoroughly applying logical or statistical techniques to illustrate, summarize, and *assess* the data. The approach to data analysis will depend on whether the research is quantitative or qualitative.

1. **Quantitative data analysis** – Advanced statistical techniques are essential for conducting sophisticated analyses that go beyond basic descriptive statistics, enabling researchers and analysts to uncover deeper insights and complex patterns in data. One such method is factor analysis, which is used to identify underlying relationships between observed variables by grouping them into a smaller set of latent factors. This technique is particularly useful in psychological and social sciences, where common hidden factors may influence multiple variables. Another powerful method is cluster analysis, which involves grouping similar data points based on shared characteristics. This technique is widely applied in market

segmentation, customer profiling, and pattern recognition, helping businesses and researchers identify meaningful clusters within large datasets. Additionally, structural equation modeling (SEM) is an advanced technique that allows for testing and estimating relationships between multiple variables, often incorporating both direct and indirect effects. SEM is particularly useful in fields like economics, psychology, and social sciences, where complex relationships between observed and latent variables need to be examined. By leveraging these advanced statistical techniques, analysts can enhance the accuracy of their findings and make more informed, data-driven decisions.

2. **Qualitative data analysis** – Qualitative data investigation engages the interpretation of non-arithmetic data, such as audio, images, or text, to understand meanings, themes, and patterns.

Thematic analysis – Thematic analysis is a qualitative research method used to identify, analyze, and interpret patterns or themes within data. The process begins with adaptation, where researchers engage with the data by understanding and re-reading it to develop a deep understanding. Next, coding involves assigning initial labels to specific segments of the data to categorize key pieces of information. Once coding is complete, researchers move on to searching for themes, grouping related codes into broader, meaningful themes that capture the core aspects of the data. These themes are then reviewed, refined, and validated to confirm that they correctly represent the dataset. In the defining themes stage, each theme is clearly described and labeled to convey its meaning effectively. Finally, the reporting phase involves presenting the themes in a structured and coherent manner, often in the form of a narrative or analysis that highlights key findings. By systematically organizing and interpreting qualitative data, thematic analysis provides valuable insights into patterns, perspectives, and meanings within research studies.

Narrative analysis – Narrative analysis is a qualitative research method that focuses on examining the structure and content of stories to understand how individuals interpret and make sense of their experiences. The process begins with identifying narratives, where researchers extract personal accounts, interviews, or textual data that contain meaningful stories. Next, analyzing structure involves examining key elements such as the plot, characters, setting, and sequence of events to understand how the story is constructed and organized. Finally, interpreting meaning delves deeper into the underlying messages, themes, and emotions within the narratives, helping researchers uncover the ways individuals construct their identities, experiences, and social realities. By studying narratives in this structured manner, researchers gain valuable insights into human perception, cultural influences, and the ways people communicate and share their lived experiences.

Grounded theory – Grounded theory is a research methodology that focuses on developing a theory directly from the data itself. It is particularly useful for exploratory research where existing theories may not fully explain the phenomena being studied. The process begins with open coding, where raw data is broken down into distinct parts, and key concepts are identified. This step allows researchers to recognize patterns and meaningful segments within the data. Next, axial coding involves reassembling the data by linking related codes to form broader categories, helping to establish relationships and connections between different concepts. Finally, selective coding is used to integrate and refine these categories, ultimately leading to the development of a cohesive theoretical framework that explains the underlying patterns observed in the data. By systematically analyzing qualitative data in this structured manner, grounded theory enables researchers to generate new insights and build theories that are deeply rooted in empirical evidence.

1.5 Statistical Significance of Analysis

Statistical significance is a foundational concept in research that helps determine whether findings are likely to be genuine or due to chance. However, it should be interpreted carefully, considering the context, effect size, and practical significance of the results. Researchers should report findings transparently, acknowledging the limitations of statistical significance and complementing it with other relevant metrics. The point-wise algorithm for the statistical significance of data is:

1. **Definition and importance** – Statistical significance is a key concept in research, determining whether observed results are due to genuine effects or random chance. It helps researchers assess the reliability of their findings and supports data-driven decision-making. Statistical significance is a cornerstone of research methodology, acting as a critical measure for determining whether the outcomes of a study reflect genuine effects or are simply the result of random fluctuations.

2. **Role in hypothesis testing** – Hypothesis testing relies on statistical significance to evaluate whether a study's results can be generalized to a broader population. This concept is integral to hypothesis testing, a process in which researchers strive to infer broader population characteristics based on observations from a sample. Through the evaluation of statistical significance, researchers estimate the probability that the patterns, trends, or differences identified in their data are not accidental but rather indicative of real relationships, effects, or phenomena. This assessment enables researchers to draw robust conclusions, minimize uncertainty, and provide evidence-based insights.

3. **Understanding p-values** – A p-value is a numerical measure that helps determine statistical significance. A commonly used threshold is 0.05, meaning that there is only a 5% chance that the observed results occurred randomly. However, lower p-values strengthen the confidence in findings. Understanding statistical significance involves not only interpreting numerical thresholds, such as p-values, but also considering the practical implications of findings within the study's context. While achieving statistical significance is often a goal in research, it is equally important to ensure that the results have substantive relevance and are not overemphasized simply due to meeting mathematical criteria. A low probability (p-value) indicates that the observed effects are unlikely to have occurred by chance.

4. **Avoid misinterpretation** – Statistical significance does not always imply practical significance. A study may yield significant results but have a small effect size, making the findings less meaningful in real-world applications. Researchers should consider both aspects when concluding.

5. **Limitations and considerations** – While statistical significance is valuable, it has limitations, such as dependence on sample size and data variability. A large sample can produce statistically significant results even for minor differences, emphasizing the need to interpret results in context.

6. **Best practices in reporting** – Researchers should transparently report statistical significance along with effect sizes, confidence intervals, and real-world implications. This ensures a balanced interpretation of results and prevents misleading conclusions.

This detailed exploration of statistical significance delves into its theoretical underpinnings, applications, and limitations, offering a comprehensive understanding of why it remains a pivotal tool in scientific inquiry.

1.6 Fundamentals of Sampling

Sampling is a key aspect of research, involving the selection of a subdivision from a greater population to make inferences about the whole. Proper sampling techniques confirm that the model correctly represents the population, enhancing the validity and generalizability of research findings. The population refers to the entire group of interest, while the sample is a smaller, representative group. Sampling is crucial due to practicality (cost and time constraints), efficiency, and the ability to generalize findings.

Key concepts include the sampling frame, sampling error, bias, and precision. Sampling methods fall into two main categories: probability sampling, where every element has a known chance of selection, and non-probability sampling, which doesn't guarantee equal chances for all elements. Methods like simple random sampling, stratified sampling, and cluster sampling are common in probability sampling, while convenience and snowball sampling are often used in non-probability sampling. A detailed description of this is summarized below.

Definition and importance – Sampling is the method of choosing a subset from a larger population to analyze and draw conclusions about the entire group. It plays a crucial role in research because studying an entire population is often impractical due to time, cost, and logistical constraints. By using a representative sample, researchers can make valid inferences about the whole population without needing to study every individual. Proper sampling ensures that findings are accurate, reliable, and applicable to broader contexts.

Population versus sample – The population refers to the entire group of individuals, items, or events that researchers are interested in studying. It can be large, such as all residents of a country, or small, like employees of a specific company. A sample is a smaller, carefully chosen subset of the population used to conduct the study. A well-selected sample should accurately represent the population, minimizing errors and biases to ensure the validity of research conclusions. Determining the right sample size is essential for reliable results, with factors like population size, desired precision, and available resources influencing it. Sampling challenges include non-response, sampling bias, and ensuring generalizability. Ultimately, selecting the appropriate sampling method is crucial for obtaining valid and representative research findings.

1.6.1 Types of Sampling Methods

Sampling methods are broadly classified into two categories:

1. **Probability sampling**: Every single person in the population has a known and equal chance of being chosen. This method reduces bias and increases the likelihood that the sample represents the population accurately. Probability sampling techniques ensure fair selection and increase the reliability of findings. Common methods include the following:

   a. **Simple random sampling**: Every individual in the population has an identical and independent chance of being chosen. This method eliminates selection bias but may require a large sample size for accuracy.

   b. **Stratified sampling**: The population is divided into different subgroups (strata) based on specific characteristics (e.g., age, income level), and random samples are taken proportionally from each stratum. This improves representation and accuracy.

   c. **Cluster sampling**: The population is divided into groups (clusters), and entire clusters are randomly selected for study. This method is cost-effective and efficient for large populations but may introduce sampling errors if clusters are not truly representative.

2. **Non-probability sampling techniques**: Selection is based on convenience, judgment, or voluntary participation rather than randomization. This method is useful in exploratory research but may lead to bias and limit generalizability. It is also useful when random selection is impractical or unnecessary. Common techniques include the following:

   a. **Convenience sampling**: Participants are chosen based on their easy availability. While quick and cost-effective, this method risks selection bias.

   b. **Snowball sampling**: Existing participants recruit others they know, making it useful for studying hard-to-reach populations (e.g., drug users or rare disease patients). However, the sample may lack diversity and be non-representative.

   c. **Judgmental sampling**: The researcher selects participants based on their expertise or specific characteristics. This method is effective when studying specialized subjects but may introduce researcher bias.

1.6.2 Sampling Errors and Biases

1. **Sampling error**: Differences between the sample and the population due to chance. Larger samples reduce sampling errors.
2. **Sampling bias**: Systematic errors that occur when certain groups are over- or under-represented in the sample. Bias can result from poor sampling techniques, non-response, or misclassification. Reducing bias ensures that findings are generalizable and accurate.

1.6.3 Determining Sample Size

The appropriate sample size depends on multiple factors:

1. **Population size**: Larger populations typically require larger samples for accuracy.
2. **Margin of error**: A smaller margin of error requires a larger sample.
3. **Confidence level**: Higher confidence levels (e.g., 95%) need larger samples.
4. **Available resources**: Practical constraints such as funding and time influence sample size. A well-calculated sample size enhances the reliability and validity of research findings.

1.6.4 Challenges in Sampling

1. **Non-response bias**: Some participants may refuse to participate, leading to an unrepresentative sample.
2. **Achieving representativeness**: Ensuring the sample truly reflects the population can be difficult.
3. **Data collection constraints**: Time, budget, and logistical limitations may restrict sampling effectiveness. Researchers must use proper strategies to overcome these challenges and ensure meaningful results.

1.6.5 Application in Research

Proper sampling enhances the reliability and validity of research conclusions. It allows researchers to generalize findings from a sample to the entire population while minimizing costs and effort. Sampling is widely used in fields such as market research, medical studies, social sciences, and political surveys to derive insights that influence decision-making and policy formulation.

1.7 Interpretation and Report Writing

Interpretations are essential skills for effectively analyzing and communicating information. Interpretation involves examining data, identifying patterns, and drawing meaningful conclusions while ensuring objectivity and accuracy. It requires critical thinking to relate findings to the research objectives or real-world applications. It also plays a crucial role in the effective communication of information, particularly in research, business, and academic fields. It requires a deep understanding of the subject matter, critical thinking, and the ability to relate findings. Proper interpretation ensures that data is not misrepresented or misused, allowing for accurate conclusions that can influence decision-making. This process involves comparing results with research objectives, identifying limitations, and making logical inferences based on evidence. Once data has been interpreted, the next step is to present the findings in a structured and comprehensible manner through report writing. The schematic for interpretation and report writing is shown in Figure 1.5.

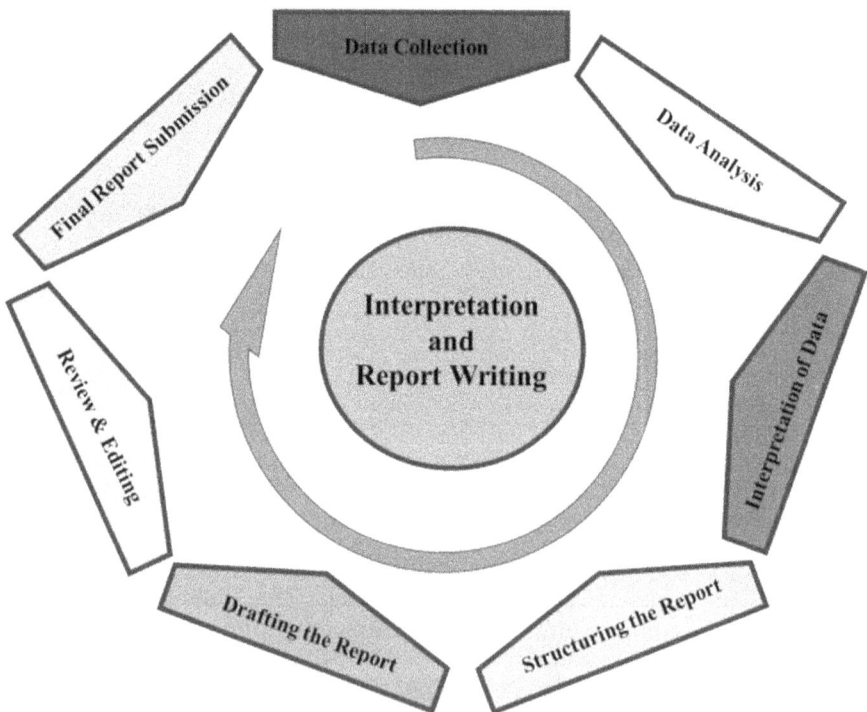

FIGURE 1.5
Flow process representing interpretation and report writing.

1.7.1 Data Collection

Gather relevant data through research, surveys, or experiments. Data collection is the initial and crucial phase of report writing, where relevant information is gathered from various sources. This may include primary data obtained through surveys, interviews, observations, and experiments or secondary data collected from books, research papers, reports, and online databases. The accuracy and reliability of the report largely depend on the quality of the collected data. Proper planning is essential to ensure that the data aligns with the research objectives. Organizing and storing the data systematically help facilitate easy access and analysis in later stages.

1.7.2 Data Analysis

Process and analyze data to identify patterns, trends, and relationships. Once the data is collected, it undergoes analysis to extract meaningful insights. This involves organizing the data, identifying patterns, and applying statistical or qualitative methods to interpret findings. Analytical tools such as Excel, SPSS, or Python may be used to process large datasets for better accuracy. The goal is to detect trends, relationships, and significant factors that contribute to the study's purpose. Errors and inconsistencies in the data must be addressed at this stage to maintain credibility. A well-conducted analysis helps transform raw data into valuable information that supports decision-making.

1.7.3 Interpretation of Data

Data interpretation involves making sense of the analyzed information by linking it to research questions or objectives. This step helps in drawing conclusions and identifying the implications of the findings. It requires critical thinking to assess whether the results align with expectations or reveal unexpected trends. Interpretation also involves comparing findings with existing literature or previous studies to establish their significance. Any limitations or constraints in the data must be acknowledged to provide a transparent and unbiased evaluation. Effective interpretation strengthens the report's credibility and impact.

1.7.4 Structuring the Report

Organize information into a logical format (Introduction, Methodology, Findings, Discussion, Conclusion, and Recommendations). A well-structured report enhances readability and ensures a logical flow of information. The report is typically divided into sections such as the introduction, methodology, results, discussion, conclusion, and recommendations. Each section serves a specific purpose, guiding the reader through the research process systematically. Proper formatting, headings, and subheadings help organize content clearly. A structured approach ensures that the report is coherent and that all relevant aspects are covered without redundancy. Effective structuring allows readers to navigate the report easily and grasp key insights.

1.7.5 Drafting the Report

Write the report with clear, concise, and factual content. The drafting stage involves putting all the collected and analyzed information into written form. A clear and professional writing style should be maintained to convey ideas effectively. The introduction sets the context, the methodology explains how the data was collected, and the findings present analyzed results. The discussion interprets the results, while the conclusion summarizes key insights. Citations and references should be included to acknowledge sources and avoid plagiarism. A well-drafted report serves as the foundation for the final version, requiring further refinement through review and editing.

1.7.6 Review and Editing

Proofread, revise, and refine the report for clarity and correctness. Reviewing and editing are critical steps to enhance the clarity, accuracy, and overall quality of the report. This process involves checking for grammatical errors, logical inconsistencies, and formatting issues. Feedback from peers, mentors, or supervisors can help identify areas for improvement. Editing also ensures that the report follows the required guidelines and maintains a professional tone. Any unnecessary details or redundant content should be removed to improve readability. A thorough review guarantees that the final report is polished and free of errors.

1.7.7 Final Report Submission

Present the final document to stakeholders, decision-makers, or the intended audience. The final stage of report writing is the submission of the completed document. Before submission, a last round of proofreading should be done to ensure accuracy and adherence to the required format. The report should be presented in a well-organized manner, with a title page, table of contents, and properly labeled sections. Submission may be in digital or printed format, depending on institutional or organizational requirements. Ensuring timely submission is crucial to meeting deadlines and fulfilling the intended purpose of the report. A well-prepared final report reflects professionalism and thorough research.

Once the data is interpreted, report writing organizes these insights into a structured format, making them accessible to readers. A well-written report typically includes an introduction, methodology, findings, discussion, conclusion, and recommendations. Clarity, conciseness, logical flow, and factual accuracy are crucial for an effective report. Together, interpretation and report writing facilitate informed decision-making and knowledge sharing across various fields.

A well-structured report is essential for conveying information clearly to its intended audience. It typically follows a systematic format, including an introduction that provides background information and objectives, a methodology section that describes how data was collected and analyzed, a findings section that presents key results, a discussion that interprets these results in context, and a conclusion that summarizes insights while offering recommendations for future action. Effective report writing requires clarity, conciseness, and logical flow to ensure that the information is easily understood. Additionally, objectivity and accuracy are crucial in preventing misinformation or bias. The use of tables, charts, and visual aids can enhance the readability and comprehension of reports, making complex data more accessible. Whether in academic research, corporate analysis, or technical studies, the ability to interpret findings accurately and communicate them effectively through reports is essential for informed decision-making and knowledge dissemination. In essence, interpretation and report writing are interconnected processes that transform raw data into valuable information, enabling individuals and organizations to draw meaningful conclusions and take appropriate actions.

Theoretical Questions and Problems

1. What is a research methodology, and why is it important in conducting research?
2. How does the choice of research methodology influence the research outcomes?
3. What are the key differences between qualitative, quantitative, and mixed-methods research?
4. Why is objectivity important in quantitative research, and how is it maintained?
5. What are the characteristics of a well-defined research objective?
6. How do descriptive, exploratory, explanatory, and evaluative research objectives differ?
7. In what ways does the research objective influence the choice of research design?

8. What are the key components of research design, and how do they contribute to a study's reliability?

9. What are the advantages and limitations of using surveys/questionnaires in quantitative research?

10. How does qualitative observation differ from quantitative observation in data collection?

11. What are the strengths and challenges of using mixed-methods research?

12. How does secondary data analysis contribute to research, and what are its limitations?

13. What is statistical significance, and why is it important in research methodology?

14. How does statistical significance help distinguish real effects from random fluctuations in research findings?

15. What role does hypothesis testing play in evaluating statistical significance?

16. Why is it important to consider both statistical significance and practical relevance in research findings?

17. What are some common misconceptions about statistical significance in data analysis?

18. How can researchers ensure the transparent reporting of statistically significant findings?

2

Literature Survey Methodology

2.1 Introduction

A literature survey, also known as a literature review, is a systematic examination of existing scholarly work related to a specific research problem or field of interest. Its objective is to evaluate and synthesize published research to understand the current state of knowledge, highlight research gaps, and identify opportunities for further investigation. It also provides comprehensive insights into the topic, reveals limitations of prior studies, and helps avoid duplication of research, while developing a structured narrative that integrates insights from multiple sources.

2.2 Identification of Broad Research Field

Identifying a broad research area is a pivotal first step in the research process, as it lays the groundwork for formulating specific research questions, objectives, and methodologies. This broader domain represents a general field of study that aligns with the researcher's academic background, personal interests, technical expertise, and responsiveness to emerging global or industrial challenges. The selection of a broad research area is typically informed by a self-assessment of the researcher's competencies and passions, combined with a critical review of existing literature, analysis of current technological advancements, and evaluation of societal or industrial needs. Active engagement with academic advisors, industry professionals, and interdisciplinary teams further enhances the relevance and precision of this selection process.

A thorough understanding of research gaps, limitations in current research, and opportunities for innovation within the field is essential to ensure that the chosen area holds both academic significance and practical value. Once a broad domain, such as Advanced Manufacturing, Biomaterials, or Sustainable Energy Systems, has been identified, the researcher can then narrow their focus

DOI: 10.1201/9781003664871-2

to a specific niche problem. This refined scope enables targeted investigation, facilitates solution-oriented research, and ultimately contributes to impactful and meaningful outcomes in both academic and real-world contexts.

In the field of science and technology especially in healthcare-oriented engineering sciences , indentifying of a broad research area is a crucial first step in addressing complex medical challenges through innovative material-based solutions. This stage guides the development of impactful research problems that advance biomedical technologies. A comprehensive understanding of the researcher's academic background expertise existing scientific literature, of emerging research trends, and unmet clinical needs are essential. Key considerations for selecting of a relevant research area include the mechanical integrity, biocompatibility, and biodegradability of materials, along with their safe and functional interaction with biological systems. These factors are central to designing materials for medical devices, implants, tissue scaffolds, and drug delivery systems. By identifying research with these priorities, researchers can develop targeted solutions that bridge the gap between materials science and clinical application.

For example, a broad research area such as "Advanced Biomaterials for Hard Tissue Regeneration" can encompass several focused directions, including biodegradable composites, surface modification techniques, biocompatibility studies, and personalized implants. Interdisciplinary collaboration and awareness of global health challenges also play an essential role in identifying high-impact research directions. Therefore, the identification of a broad research area involves structured steps that align researcher's expertise and support the development of a focused and relevant research agenda. The flowchart illustrating the key stages involved in this process is presented in Figure 2.1.

The initial step involves identifying subject areas or disciplines that align with the researcher's interests and motivations. It includes reflecting on personal passions, academic strengths, and prior exposure to relevant research fields, which collectively help in narrowing down potential domains into feasible output.

2.3 Identification of Landmark Articles/Review Articles

After identification of a broader research area, finding landmark articles or review articles in the identified domain is a crucial step for conducting quality research, especially when entering a new field or refining expert research direction. There are key steps to identify such impactful articles, such as to use academic databases strategically. The research articles can be easily searched through research platforms like Google Scholar, Scopus, Web of Science, and Research Gate, where specific articles can be found using keywords and sorting results. Furthermore, selecting the most suitable articles

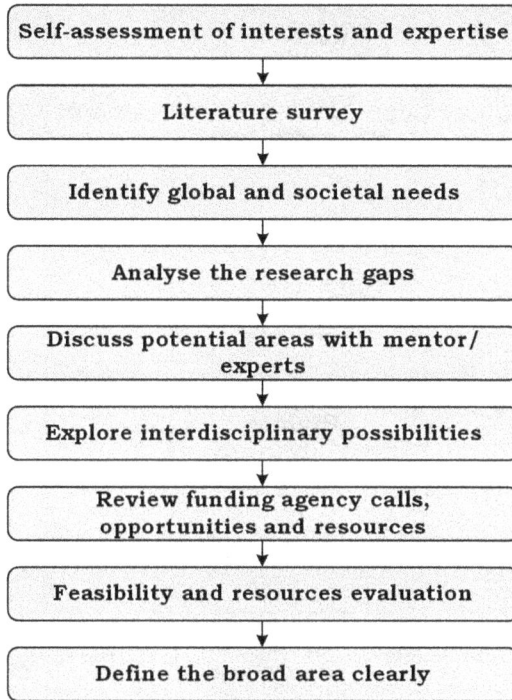

FIGURE 2.1
A flow chart of the steps to be followed in the identification of a broad research area before conducting research.

from searched articles is again a challenging task, however, filters can help to shortlist the highly cited papers and the most appropriate review articles. Landmark articles are typically highly cited, indicating they have significantly influenced the field. To find out the review articles, some keywords like review, state of the art, systematic review, and comprehensive review, along with the main topic, are used in various search platforms, including *ScienceDirect, IEEE Xplore, SpringerLink, etc.* A good review article cites several seminal or original research papers; track these references to discover landmark studies.

In addition, articles published in reputable and high-impact journals should be kept at top priority; examples of such journals for research on biomaterials are *Nature, Science, Advanced Materials, Biomaterials, Acta Biomaterialia, Journal of Biomedical Materials Research, etc.* The identification process of a landmark or review article of any domain is displayed through a flowchart in Figure 2.2. The selection of a journal and then filtering out the research articles play a crucial role.

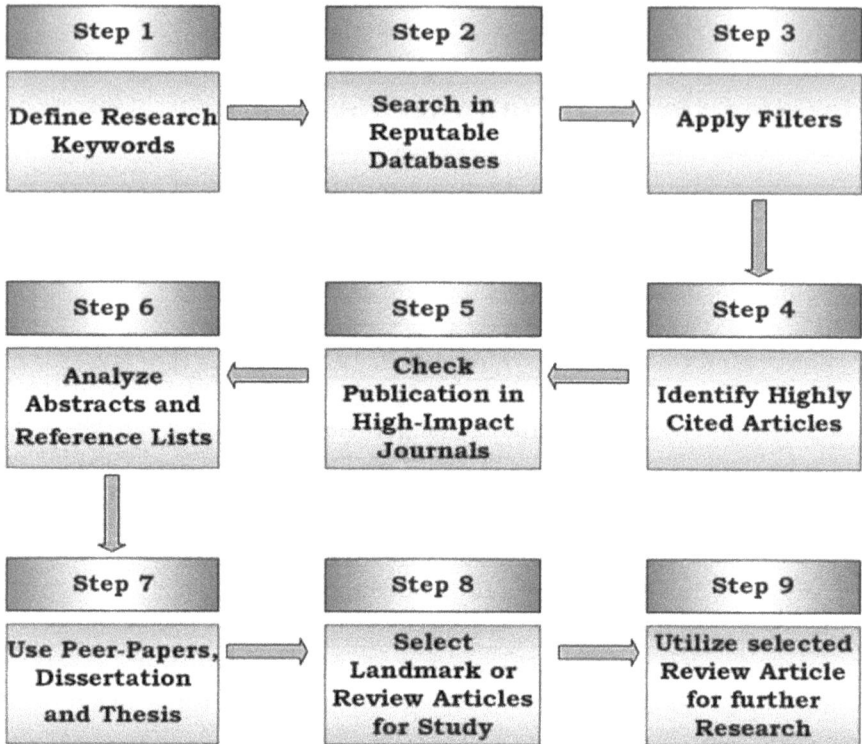

Step 1	Step 2	Step 3
Define Research Keywords	Search in Reputable Databases	Apply Filters

Step 6	Step 5	Step 4
Analyze Abstracts and Reference Lists	Check Publication in High-Impact Journals	Identify Highly Cited Articles

Step 7	Step 8	Step 9
Use Peer-Papers, Dissertation and Thesis	Select Landmark or Review Articles for Study	Utilize selected Review Article for further Research

FIGURE 2.2
A flowchart shows the identification process of a landmark or review article in new research.

2.4 Identification of Some Potential Research Problems

A thorough literature review is an essential key step for the identification of a potential research problem. To finalize a potential research problem is to deeply read recent papers, last three to five years in the targeted area. During the literature survey, it should be focused on the sections of "Future Work," "Limitations," and "Results and Discussion," where researchers can often mention research gaps that still need to be solved. In addition, different research articles on similar problems also provide research gaps and unresolved issues, along with various justifications and solutions. Along with current trends, present requirements, and future perception should also be considered to explore the emerging trends in new materials, new methods, and new needs.

To find out the latest trends in scientific research, popular journals, patents, and even conference proceedings should be included in the thorough literature review. It also includes discussion with experts, professors, industry professionals, lab colleagues, etc., to know what challenges they faced. Their practical observations are beneficial in finding the challenges, existing assumptions, and better ways to approach the challenges. In addition, a proper planing of a small problem can lead to high impact in research.

2.5 Preliminary Experimental/Simulation Analysis

This section aims to perform an initial investigation, either experimentally or through simulations, to understand the basic behavior, performance, or viability of the proposed research idea before fully committing to a large-scale study. The following steps are to be followed before conducting preliminary experiments or simulation analysis:

1. Define your objectives clearly.
2. Design some experiments or simulations.
3. Conduct the experiments/simulations.
4. Collect the experimental/simulation data.
5. Analyze the collected data and find the approximate trends.
6. Modify the experiments as per the analysis of collected data and trends.

To define objectives clearly, it is required to find out the answers to a few basic questions like what specific aspect do you want to test or simulate? What types of process parameters are needed to select and optimize in preliminary experimentation/simulation? and What properties/characteristics are required to test from the application point of view? For example, if the targeted research area is development of artificial biomaterial then first step is to find out the raw materials, their compositions, processing technique, simulation study, process parameters, desired properties of expected biomaterial (mechanical strength, biological compatibility, etc.), and their testing to analyze the properties as per applications. Furthermore, looking for proof of concept and exploring the feasibility of materials, processing techniques, and process parameters optimization. Based on results, refine methods, models, or hypotheses for the full study and refine the preliminary experimentation for final processing and testing. To simulate stress distribution and load variation in various components are carried out using finite element analysis (FEA) and observe that high stress concentrations near the neck region, which lead to alterations in the design, necessitate modifications to reduce localized stress before physical prototyping. It is summarized that preliminary experimental/simulation analysis provides early evidence supporting the feasibility of the research problem and offers critical insights for refining the main study.

2.6 Feasibility Study of Various Research Problems

A feasibility study is an evaluation of how practical and possible it is to carry out a research project. Before investing time, money, and effort, researchers ask questions such as: Is this research problem feasible? Is it possible to access the technology, methods, tools, and expertise needed? Is the research methodology practical? For example, a study requiring brain magnetic resonance imaging (MRI) scans isn't feasible if the MRI equipment is not able to access. In addition, economic feasibility is also an important attribute in the feasibility check of any research project, as is the focused research problem, able to be completed under the available budget or grants? It includes the costs of equipment, travel, salaries, etc., realistic comparison to the expected benefits. After that, operational feasibility should also be checked to ensure that the project timeline, team, and organizational structure are enough to conduct the project and achieve the proposed objectives.

In addition, ethical approvals are also crucial for conducting any research project to avoid copyright and plagiarism. It includes legal and ethical feasibility, where it is focused that the research study complies with regulations like ethics board approval, participant consent, and data protection. Time also plays an important role in understanding the feasibility of completing any research within the available time frame and deciding whether the timeline is realistic for data collection, analysis, and publication. At last, scope feasibility decides the scope of the research problem, like the problem is narrow enough to be addressed properly or not.

2.7 Procedure for Defining Research Problems

Defining a good research problem is crucial because it sets the entire direction of the research study. A poorly defined problem leads to confusion, wasted effort, resources, and weak results. A step-by-step procedure should be followed to define the research problem. The first step is to identify a broad research area in which to choose a general field that belongs to the area of interest or where there's a known need for new knowledge. In addition, researchers should think about global challenges, gaps in industry, or their own academic strengths. For example like biodegradable materials, additive manufacturing in healthcare, artificial intelligence (AI) for drug discovery, etc., are in current trends. Figure 2.3 shows the various steps (clockwise direction) that are used to follow in defining the research problem.

After finalizing the broad research area, a thorough literature review is required to be conducted, which includes existing research papers, patents, review articles, and case studies. A thorough literature survey assists in understanding the area and identifying research gaps, inconsistencies, unanswered questions, or controversial findings and opportunities for improvement. A detailed literature review helps in narrowing down the broader research area into a specific topic after finding out the research gaps. It focuses on a specific

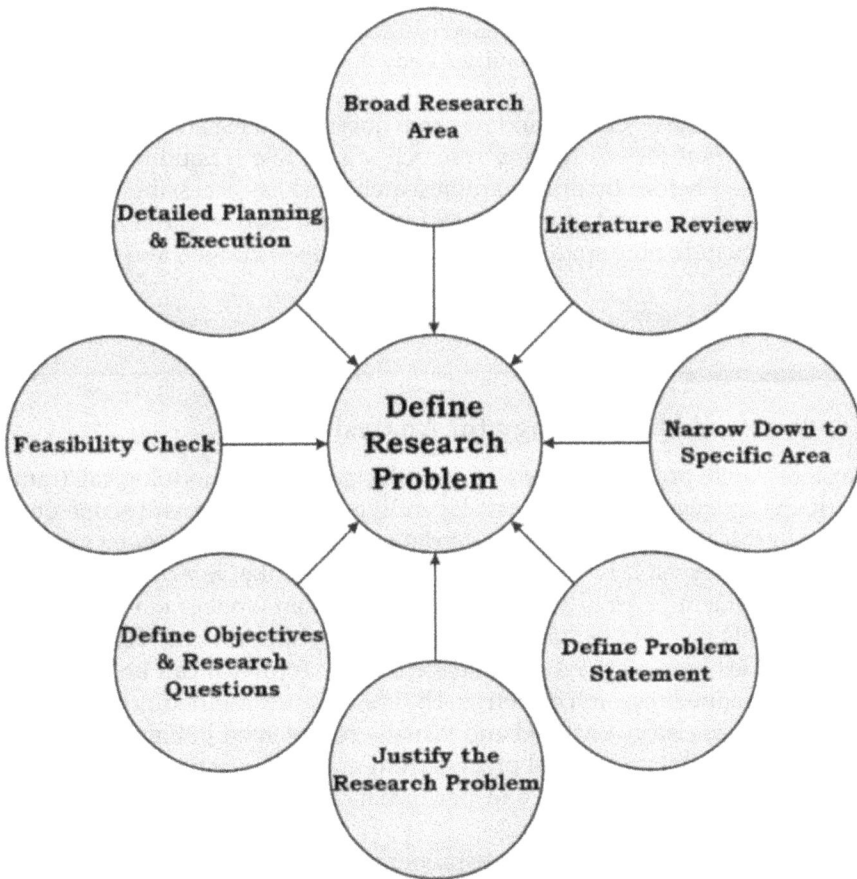

FIGURE 2.3
Various steps to be followed to define the research problem statement.

issue or challenge within the broad area. Furthermore, research gaps state the problem undoubtedly and write a clear, concise statement of the research problem, including what is to be investigated, why it is important, and what gaps are to be filled. For example:

> Current titanium orthopedic stems cause stress shielding in patients, leading to implant failure. Hence, a research study is needed to investigate the design of biodegradable composite stems to enhance bone regeneration and reduce stress shielding. Further, to justify the research problem, scientific importance, industrial or social relevance, and practical applications, show that it's a real problem worth solving. Next step is to define objectives and research questions by breaking the problem into specific objectives and frame research questions that guide the study in the specific direction to achieve the proposed objectives; for example "to develop a 3D-printed biodegradable orthopedic stem, to analyze its mechanical properties and degradation rate, and to analyse the degradation kinetics of biomaterial in simulated body fluid conditions."

After defining the objectives and research questions, a feasibility check is one of the important factors for any research study. Test feasibility is essential to be checked before finalizing any research work, as accessing the needed resources is also set limitation in achieving the proposed research work. These resources include time span, equipment availability, budget, and ethics.

2.8 Proposed Methodology for Analysis

Most research problems follow a broadly similar methodological framework; the specific steps may vary depending on the nature and scope of the study. In this context, the process of defining the methodology or methods for scientific research is illustrated using a representative example, such as the development of an orthopedic stem made from biodegradable composites using 3D printing technology for healthcare applications. This example demonstrates how a systematic and structured approach can be applied to address a defined research objective. The step-by-step methodology adopted in this study is clearly outlined and visually represented in Figure 2.4, offering insight into the sequential process from material selection and design to fabrication, characterization, and performance evaluation tailored to meet specific biomedical requirements.

Material selection and composite preparation are the first steps in the mentioned research problem, where biodegradable polymers/metals/ceramics and fillers are to be selected. To decide the composition of composites is one of the vital steps that will be decided after a thorough literature review.

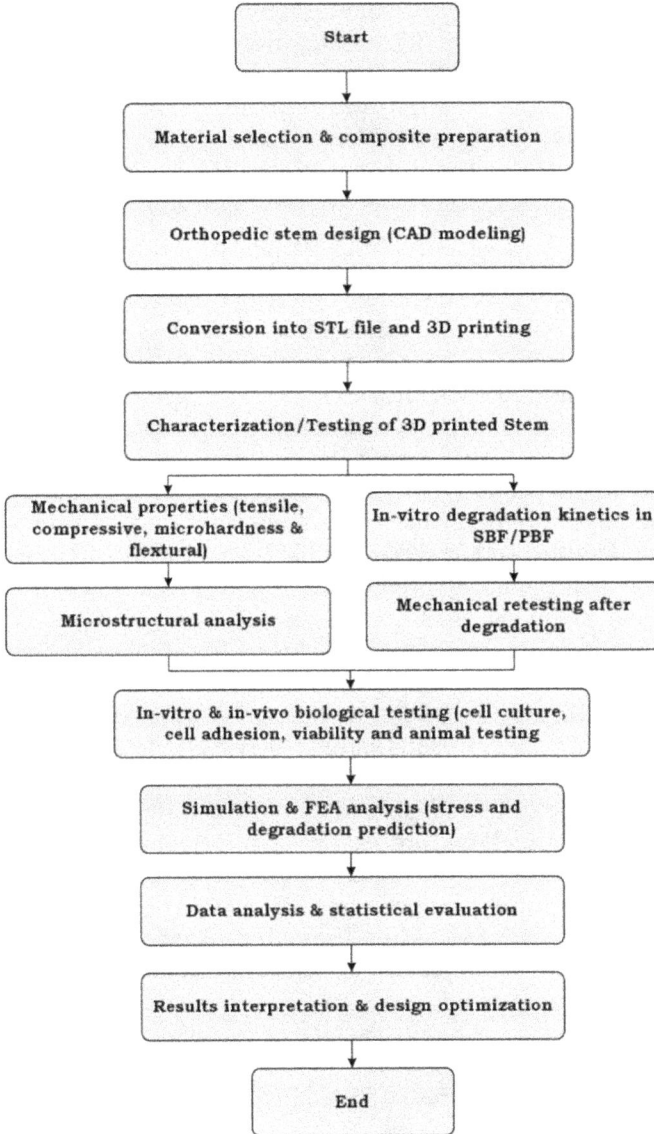

FIGURE 2.4
Methodology of development of orthopedic stem of biodegradable composite through 3D printing for healthcare application.

Next, the homogenized polymer-filler mixtures using melt blending or solvent casting methods before 3D printing need to be made. CAD model of the stem is designed before 3D printing, as the design of the stem geometry incorporates internal porous structures optimized for mechanical strength and bone ingrowth. The designed CAD model is to be converted into STL (stereolithography) format, and then select the 3D printing process like fused deposition modeling (FDM) or selective laser sintering (SLS) based on material type. The 3D printed stem is used for mechanical and degradation tests to measure mechanical properties along with degradation kinetics in simulated body fluid (SBF) at 37°C for 1, 7, 14, 28, and 56 days (or more depending on applications). This test also depicts the weight loss and helps in understanding the degradation rate, pH measurement, and measurement of concentration of released ions in SBF.

Furthermore, integration of mechanical and degradation properties is also significant for biodegradable composites, as this testing can only estimate the performance in the actual working environment. Microstructural and morphological analysis are performed to examine surface morphology, porosity, degradation patterns, and analyze internal pore structure non-destructively. After getting the results of preliminary experimentation, simulation and computational modeling are to be performed using FEA, which analyzes the stress distribution within the printed stem under physiological loads. It helps in the prediction of weak points and optimizes the design. It also optimizes the degradation rate and simulates gradual material loss to predict long-term behavior using various software like ANSYS, Abaqus, SolidWorks Simulation, or similar platforms.

In addition, biological analysis can be done through *in vitro* biocompatibility testing, where cell culture studies (e.g., osteoblasts or mesenchymal stem cells) are used to test cell adhesion, proliferation, and viability. These tests should be performed as per standards like MTT (3-(4, 5-dimethylthiazol-2-yl)-2, 5-diphenyltetrazolium bromide)) assay and Live/Dead staining. At the end, the obtained results and data from different tests are to be analyzed and interpreted into a statistical format. The interpretation of data is carried out using ANOVA, t-tests, or regression analysis to compare mechanical properties, degradation rates, biological responses, etc. Finally, interpreted data is used for research publications, patents, and other scientific reports.

Theoretical Questions and Problems

1. How do we define the goals and objectives for the literature review?
2. What theoretical perspectives or concepts are relevant to the research question?
3. What are the initial assumptions or hypotheses about the research?
4. How will the literature review help to refine or test the initial assumptions?
5. How do we identify the strengths and weaknesses of the existing research on the targeted topic?
6. How do we identify the main findings and conclusions of the reviewed literature?
7. What gaps or debates exist in the current research? How do we identify?
8. How do the review articles contribute to enhancing the understanding of the research problem?
9. How do we define the potential benefits of the targeted research to the field or society?
10. Are there specific geographic, demographic, or temporal limitations to the focused research?
11. How do you assess that the focused research question feasible and achievable within the constraints of resources and time?

3

Classification of Research Approaches

3.1 Introduction

Let us suppose we want to design a beam of a bridge for a particular load. The dimensions of the beam depend on the strength of the material used to fabricate the beam. Thus, evaluating and assessing the beam material's response to load are vital for designing the beam. Experimental, simulation, and analytical are the three main approaches that could be used to evaluate, analyze, and assessment of the material's response. The total deformation corresponding to a particular load could be estimated using all three methods mentioned above. These three methods could be used for any problems you may think of. In general, the experimental method is more accurate than analytical, followed by simulation. The accuracy of the simulation method depends on the assumptions, length, and time scale. We will see a detailed description of these methods in this chapter.

3.2 Experimental Approach

This is more accurate among the three methods because it considers the effect of all the possible factors and represents the actual response of the materials. The area of interest in the experimental analysis method should be sufficient to have statistical significance and to represent the response of the large components. The schematic representation of the magnified view of the particular at different length scales is shown in Figure 3.1. Several aspects should be considered to execute this approach. Some of the important steps are as follows.

3.2.1 Statistical Significance

Several experiments need to be conducted multiple times under a particular condition to verify the consistent response of the materials and to ensure the reliability of the test method. For example, several micrographs are taken of a

DOI: 10.1201/9781003664871-3

Macroscopic 10 mm

External tractions

Microscopic (100 μm)

Grains & textures

Atomistic (1 nm)

Dislocation core

Sub-microscopic (1 μm)

Dislocations

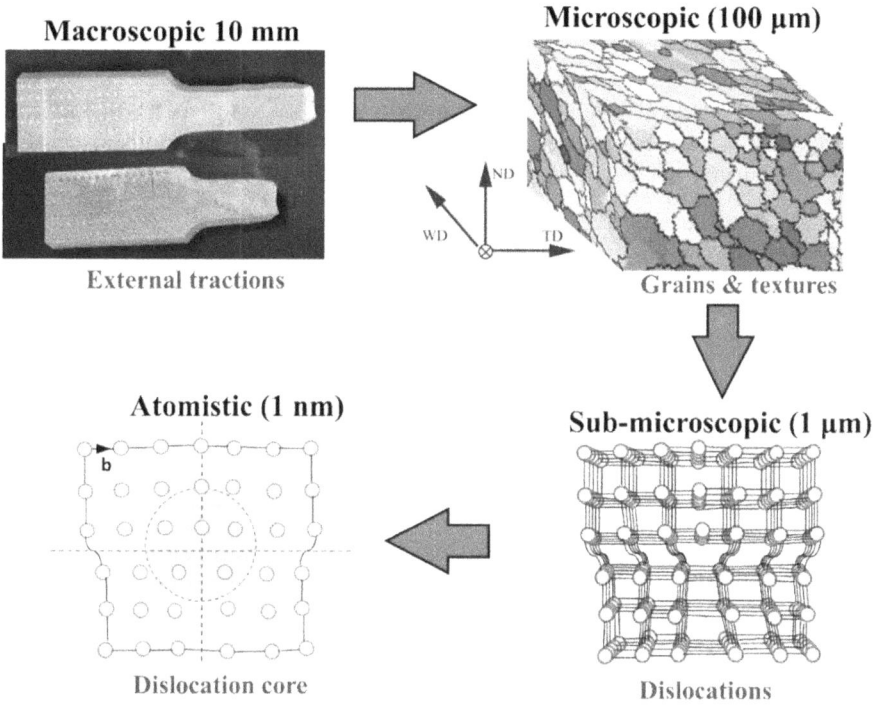

FIGURE 3.1
Schematic representation of the material response at different length scales.

sample using characterization techniques and then are analyzed; finally, representative micrographs should be presented in the report/article. This is a known fact that the area of analysis in some characterization techniques is usually less than the actual size of the deformed sample, which does not show the statistical consistency and representativeness of analysis data. Therefore, a more appropriate investigation should be performed to get the overall response of the materials. The methodology adopted for the investigation must highlight the integrity and statistical significance.

3.2.2 American Society for Testing and Materials Standards

It is a scientific and technical organization formed in 1898 for the development of standards on the characteristics and performance of materials and related knowledge. The purpose of this manual is to promote uniformity because it facilitates easier access to information and quicker understanding for the user. Deviations from the American Society for Testing and Materials (ASTM) could lead to inaccuracy in measurement and inefficiency for authors, reviewers, editors, and, eventually, the reader of these standards.

The ASTM standards in testing are globally recognized guidelines by ASTM International that define the essential procedures for testing materials such as metals and plastics. These standards define how to measure material properties such as tensile strength and elongation, guaranteeing accurate and consistent results, vital for material quality and safety in different industries. The most commonly used ASTM standards are listed as follows.

Sr. No	Standards	Description
1	ASTM A370	Tensile strength of metals
2	ASTM A938	Torsion testing of wire
3	ASTM E18	Hardness testing of metals
4	ASTM E466	High-cycle fatigue testing
5	ASTM E467	Fatigue crack growth testing
6	ASTM E399	Fracture toughness testing
7	ASTM E606	Low-cycle fatigue testing

3.2.3 Testing Method Selection

The selection of the right method depends on the nature of the research (qualitative, quantitative, or mixed), norms of the research area (clinic and pre-clinic permission), and practicalities of the methodology (constraints faced practically cannot be overlooked) and length scale (macro, micro, nano, or atomic, Figure 3.2). For selecting the appropriate testing

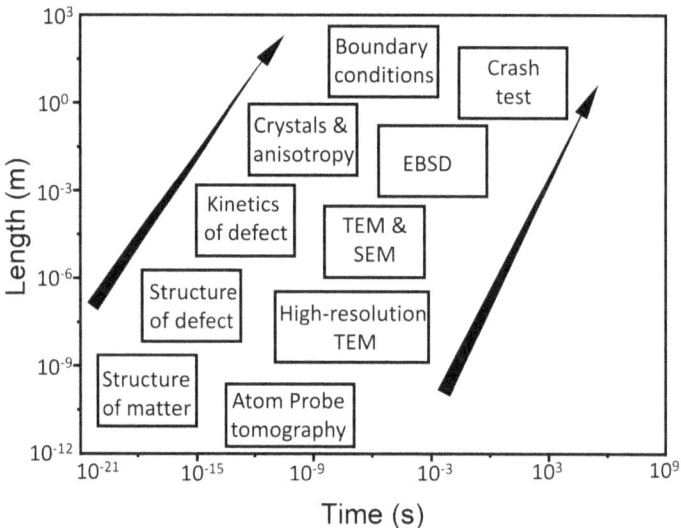

FIGURE 3.2
Selection of the characterization technique based on need and the length scale.

method, there are several steps, need to be considered, which are as follows:

1. Define the goals, objectives, and research questions.
2. Refer to pertinent research and effectively used methodology.
3. Structuring the plan and finding resources to conduct research.
4. Write the research methodology in detail and review it.

3.2.4 Feasibility Analysis

Before considering a scientific problem for your investigations, it is important to do a feasibility analysis so that you can understand the issues associated with it and how you can overcome these issues over time. There are several steps to do the feasibility analysis of the problem, some of which are as follows:

1. Objectives/research statements should be defined clearly and should be specific, measurable, and achievable in a time-bound manner.
2. The comprehensive literature review helps young researchers to define a research statement that should be novel and relevant to industrial needs. It also helps in demonstrating the feasibility of your research.
3. Consider conducting preliminary studies to validate literature findings and highlight your capabilities to pursue the research problem further. It would demonstrate the viability of your approaches and understanding to do the feasibility analysis of your research.
4. A detailed outline of your proposed methodology and research design should be comprehensive, and all the steps you will take to address your research questions must be mentioned in the methodology. If you employ multiple methods, it's crucial to justify why you've selected these specific approaches for your investigation and analysis.
5. The potential challenges and risks associated with the adopted methodology should be mentioned, and possible mitigation strategies could be presented that highlight your preparedness and enhance the feasibility of your research plan. These challenges could be related to technical limitations, ethical considerations, and personal constraints.
6. Researchers should be transparent about the time required for each phase of the research process, and a realistic timeline should be presented for each step. A well-structured timeline demonstrates

your ability to manage time effectively and enhances your chances of success.

7. Research objectives should be defined by considering the available resources and possible collaboration opportunities. The personal capability of researchers to access the available specialized equipment, the availability of funding, and partnerships with other researchers will help to complete the task within the given time frame.

8. Financial resources (a well-justified budget) required to support your research activities should be defined and mentioned clearly as to how you could be sent to the successful execution of your project.

3.3 Numerical Simulation Approach

Several numerical methods are used for the simulation of engineering phenomena. Continuum, macro, micro, meso, and atomistic are the primary classifications of the numerical methods depending on the size of the model, as shown in Figure 3.3. Constitutive modeling is a macro simulation technique, whereas crystal plasticity is the micro. Mesoscale simulations are performed for features like grain clusters or phase interfaces. Atomistic methods, such

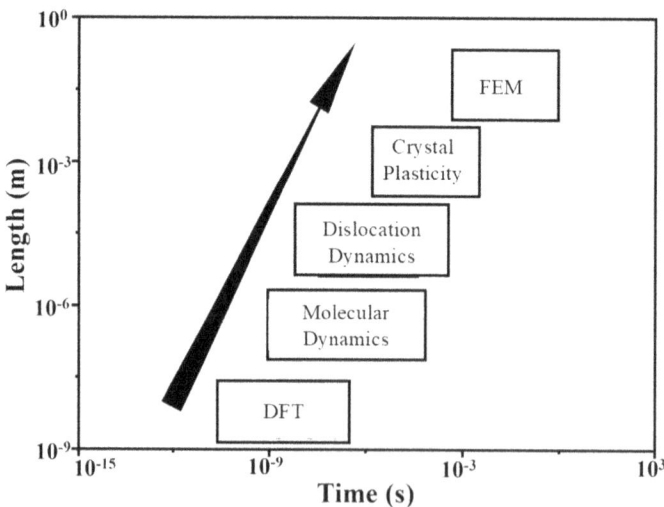

FIGURE 3.3
Various methods used for numerical simulation and analysis.

as molecular dynamics, model interactions between individual atoms, and quantum-level simulations (e.g., density functional theory) are used for electron-level analysis of materials.

Please note that the methods at a smaller scale are very localized techniques typically used for fundamental study and require significant computational resources. One needs to ensure the statistical significance of the analysis. Users could select the appropriate techniques depending on the objective of the analysis, the availability of computational resources, and the required solution. There are three main steps in the finite element simulation process. They are as follows:

1. **Preprocessing phase**: First, we discretize the problem domain into a large number of small elements through meshing, which consists of nodes and elements. After defining material properties and applying the boundary conditions with initial conditions (if required), the problem is solved. After setting up the finite element model, the governing equations are formulated. The value at the element centroid could be estimated after the analysis; then, using the shape function, the values of the different nodes are interpolated.

2. **Solution (Processing) phase**: In this stage, a set of linear or nonlinear equations is solved simultaneously to obtain the unknown variable (e.g., displacement in a stress problem) at the nodal point. After that, other required variables are calculated.

3. **Postprocessing phase**: In this phase, various field variables like stress, strain, temperature, and pressure are extracted and visualized by the user. Contour plots, vector diagrams, and various animations are generally used to get more information about the material/structure response against the loading.

3.3.1 Constitutive Models and Assumptions

Numerical simulation involves using constitutive laws (mathematical equations or empirical equations) to describe a system's behavior. Constitutive models could be used to predict the behavior of the system under different conditions with a certain level of accuracy based on the assumptions associated with the constitutive models. Users can increase the prediction accuracy and capabilities of models by modifying the available constitutive models and implementing them through writing codes. Several constitutive models could be found in the literature to use for different conditions and length scales. The main advantage of the numerical simulation approach is that researchers can explore new ideas without expensive and time-consuming experimentation.

Assumptions associated with the constitutive models or the capability of the available resources must be mentioned in the methodology so that readers can understand/interpret the purpose and applicability of the analysis.

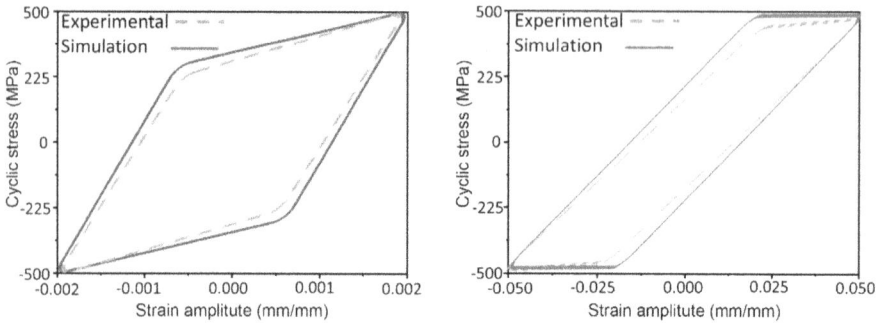

FIGURE 3.4
Schematic representation of the simulated and experimental hysteresis loops.

Assumptions could be several, depending on the complexity of the problem and the analysis. Thus, the main assumptions that could influence the analysis significantly should be mentioned.

3.3.2 Selection/Identification of Parameters

For performing the simulation, it is essential to estimate/find the material parameters that could be obtained from the experimental data or literature. Material parameters are selected based on the constitutive model to be used for simulation, and the simulation results are highly sensitive to these parameters. Once you identify and validate the parameters, the procedure of identification of the parameters must be mentioned for transparency and reproducibility so that readers can adopt the same methodology for their analysis.

3.3.3 Validation of Models and Parameters

Before relying on the simulation results and output information, the parameters and the constitutive model response should be verified. It could be done by comparing the output of the numerical simulation with the experimental results. If the experimental response matches the simulation (as shown in Figure 3.4), then it indicates the validation of the parameters and the constitutive laws.

3.3.4 Post-Processing Analysis and Visualization

Post-processing in numerical simulation involves interpreting the vast array of discrete numerical values produced by numerical simulations into visuals and insights that help engineers make informed decisions.

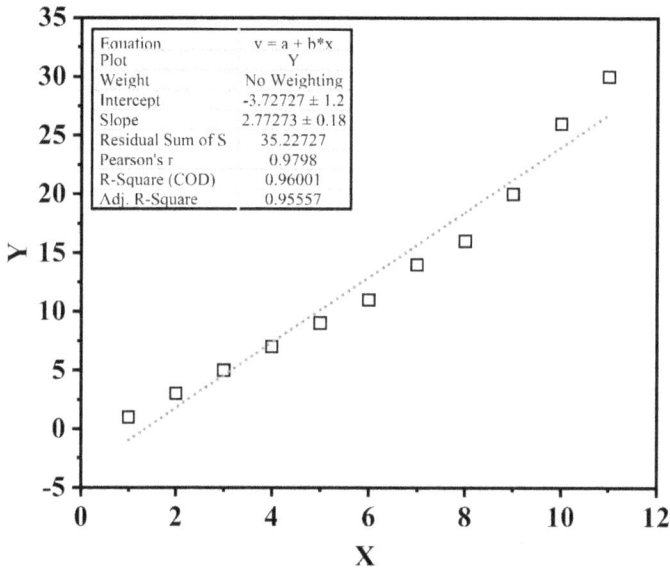

FIGURE 3.5
Schematic representation of the empirical relation.

3.4 Analytical Approaches

3.4.1 Empirical Relations

Empirical relations/equations are based on experimental data fitting and experience without a theoretical justification, as shown in Figure 3.5. They are obtained by using a set of experimental data about a particular state to predict the response to this state. They could be handy tools for design purposes and are widely used in industries. The fitting coefficient (R^2 value) decides the accuracy of the prediction of the empirical relations. In general, it should be close to the one for accurately predicting the response of the system.

3.4.2 Analytical Equations

Some equations are theoretical in the sense that they are derived from an underlying theory, like Newton's law of motion and gravitational laws. Other equations are empirical because they were selected only because they fit experimental data without a theoretical justification, like Paris Law, which is used for predicting the response of aerospace components.

In addition to these approaches/methods, recently emerging ML techniques can also be used to predict/assess the response of the material, where prediction is done based on a large input dataset by training the model. The detailed description of ML and AI is discussed in Chapter 4.

Theoretical Questions and Problems

1. What are the different methods for investigating and analyzing the research problems?
2. What are the main steps of the experimental approach and its advantages over the numerical simulation approach?
3. How do you select a proper approach for the experimental analysis? what are the main steps for selecting the proper method?
4. How to perform the feasibility study for a defined problem, and what do you understand about the statistical significance of the analysis?
5. What are the main steps that should be considered for performing the feasibility study?
6. What do you understand about the ASTM standard, and what is the significance of the ASTM standards?
7. What is the numerical simulation approach, and its steps?
8. What are the different types of numerical simulation methods based on the length scales and their advantages over the experimental approach?
9. How do we select the assumptions and parameters for performing the numerical simulation?
10. How do we validate the parameters and visualize the results of the simulation results?
11. What is the difference between the analytic equations and empirical relation? Which one should we choose for the analysis?
12. Write the main advantages, disadvantages, and limitations of the experimental approach.
13. Write the main advantages, disadvantages, and limitations of the numerical simulation approach.
14. Write the main advantages, disadvantages, and limitations of the empirical and analytical equations analysis.
15. Which is the best approach based on cost, time, quality, and feasibility?

4

Application of AI and ML for Research

4.1 Introduction to Machine Learning and Artificial Intelligence

ML and AI are playing an increasingly transformative role in modern research across diverse scientific disciplines. These technologies enable researchers to process and analyze vast amounts of complex data with greater speed, accuracy, and efficiency than traditional methods. AI and ML algorithms can identify patterns, make predictions, and optimize processes, thereby accelerating discovery and innovation. From automating data analysis and enhancing experimental design to enabling predictive modeling and personalized solutions, AI and ML are reshaping the way research is conducted. Their integration not only improves the reliability and reproducibility of results but also opens new avenues for interdisciplinary exploration and real-time decision-making.

4.1.1 Machine Learning

In the conventional programming approach, programmers write the code to transform specific tasks when certain conditions are met, and the program executes a specific action. On the other hand, ML is an automated process (with little or no human input), and it is used to train the models with the help of large data to predict behavior and further improve the prediction capability. The available ML algorithms/models could identify the patterns/ sequences in large data sets and could learn through experience from them. ML algorithms could be used to optimize and analyze large data with less time and higher efficiency. The steps (Figure 4.1) for analyzing any problems using the ML approach are as follows.

- **Data collection**: It could be in different forms such as images, text, numerical values, and videos.
- **Data preprocessing**: Need to convert the raw data into an acceptable form to give input for the ML algorithms.
- **Selection of model/algorithms**: Supervised and unsupervised models could be selected based on the task/objective of the analysis.

DOI: 10.1201/9781003664871-4

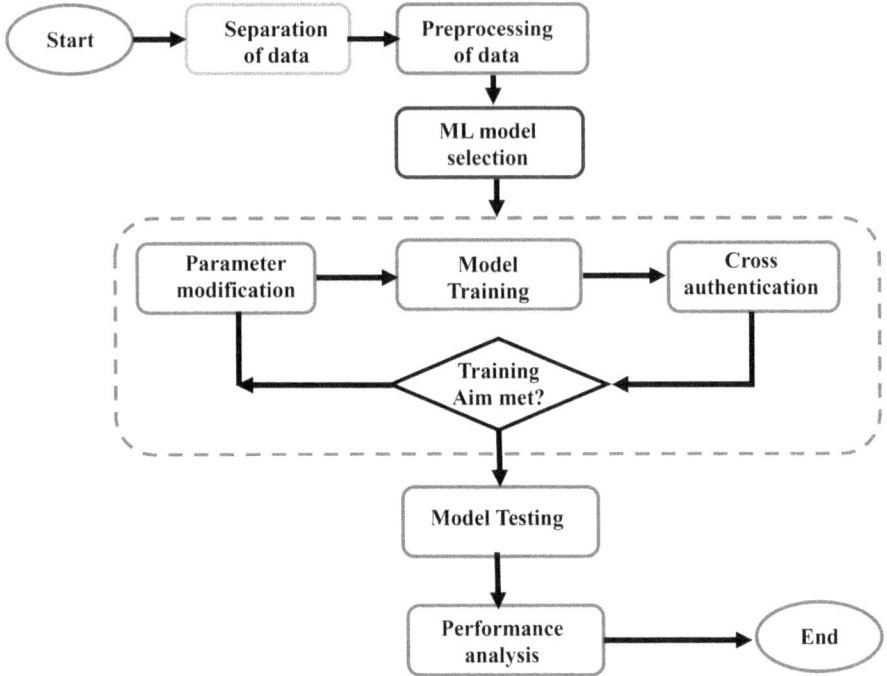

FIGURE 4.1
Flowchart of the ML process.

- **Training the model**: Approximately 80% of the data should be used to train the model, and the remaining 20% could be used to check the prediction capability of the model.
- **Response to the model**: Performance must be evaluated after training the model by comparing the response of the model and the actual response.
- **Fine-tuning**: Models/algorithms could be modified slightly to get a better output if the outcomes of the models are not matching or the error is large.
- **Prediction**: Finally, the prediction could be made, and if it is okay, then the model could be used for future analysis.

In general, ML could be classified into three categories:

1. Supervised ML
2. Unsupervised ML
3. Reinforcement ML

The principle of supervised learning is that it uses the labelled data, and the output patterns are fixed. However, unsupervised learning deals with unlabeled data where the outcomes depend on the collection of perceptions. Reinforcement learning is a bit different in that the agent interacts with the environment in discrete steps to get the output. In a nutshell, in supervised learning, the goal is to generate a formula based on input and output values. In unsupervised learning, we find an association between input values and group them. In reinforcement learning, an agent learns through delayed feedback by interacting with the environment.

4.1.2 Artificial Intelligence

It is a process of making a computer or software think intelligently like the human mind. AI is achieved by studying the patterns of the human brain and by analyzing cognitive processes such as learning, reasoning, problem-solving, perception, and language comprehension. To achieve this, the knowledge of various fields such as mathematics, biology, psychology, sociology, computer science, mechanical engineering, electrical, electronic, statistics, and neutron science is needed; thus, it is an inter-disciplinary field, as shown in Figure 4.2. To develop AI, we should know how intelligence is created in the human brain. So, Intelligence is an intangible part of

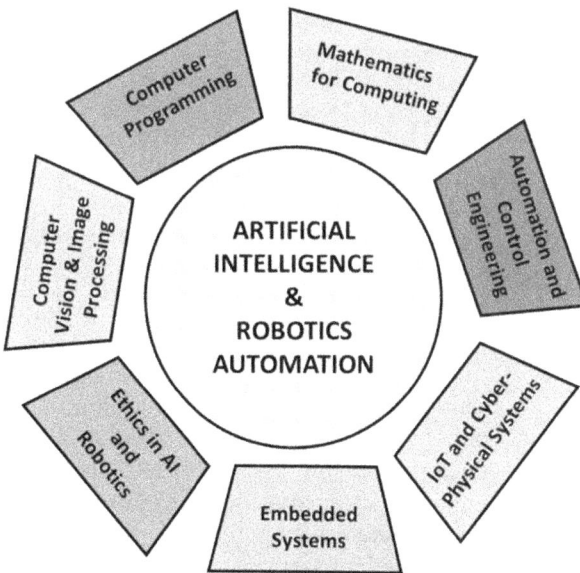

FIGURE 4.2
Schematic representation of various areas involved in the AI.

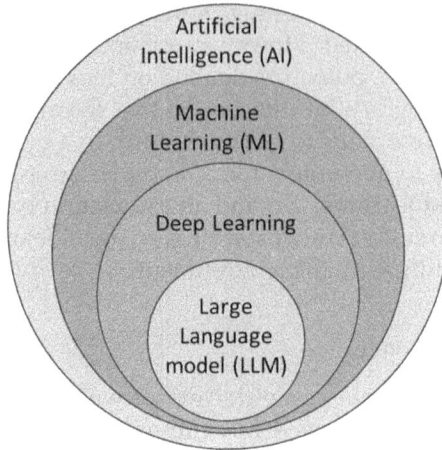

FIGURE 4.3
Schematic representation of the various fields of automation.

our brain, which is a combination of reasoning, learning, problem-solving, perception, language understanding, etc. It should be noted that ML is a subset of AI, and deep learning is a subset of ML, as schematically shown in Figure 4.3.

4.2 Need for AI and ML

AI and ML have become essential technologies in the modern world due to their transformative impact across nearly every industry and aspect of life. The primary need for AI and ML arises from the growing volume, complexity, and speed of data generation in today's digital age, which is far beyond the capacity of human processing. These technologies enable systems to learn from data, identify patterns, make predictions, and improve performance over time without being explicitly programmed for every scenario. In fields like healthcare, AI and ML are revolutionizing diagnostics, drug discovery, and patient monitoring. In finance, the power fraud detection, algorithmic trading, and customer service chatbots. In manufacturing, they drive predictive maintenance, quality control, and smart automation. Moreover, AI enhances user experiences in everyday applications such as virtual assistants, personalized recommendations, and autonomous vehicles. As the world becomes more data-driven and interconnected, the ability of machines to learn, adapt, and make intelligent decisions will be crucial for innovation, efficiency, and competitiveness. Therefore, the integration of AI and ML is not just a technological advancement, but a foundational shift that will continue to shape the future of work, education, governance, and human interaction.

4.3 Application across Disciplines

A variety of AI and ML applications are there to analyze complex datasets, make predictions, and uncover hidden patterns in different sectors of the world. The most possible uses of AI and ML are discussed below.

4.3.1 Technical Fields

- **Robotics and automation**

  Industrial robots powered by AI are used for complex tasks like assembly, packaging, welding, painting, and material handling. Collaborative robots (cobots) also work alongside human workers, increasing productivity and safety.

- **Disease diagnosis and prediction**

  AI/ML models can analyze patient data, symptoms, and medical images to detect diseases like cancer, diabetes, heart conditions, and neurological disorders with high accuracy. Tools like IBM Watson Health are used to assist doctors in diagnostics.

- **Virtual health assistants and chatbots**

  AI chatbots and virtual assistants provide 24/7 support for answering patient queries, setting appointments, sending medication reminders, and offering preliminary health advice, reducing the burden on healthcare professionals.

- **Process optimization**

  AI analyzes data from production processes to optimize workflows, energy consumption, and resource utilization. It helps in identifying bottlenecks and suggesting improvements for increased efficiency.

- **Inventory and warehouse automation**

  Autonomous robots and ML algorithms manage inventory, track shipments, and organize warehouses, reducing human labor and improving accuracy and speed in logistics.

4.3.2 Non-Technical Areas

- **Epidemic and outbreak prediction**

  AI models are used to track disease outbreaks (like COVID-19) by analyzing data from multiple sources (social media, travel data, and health records), enabling faster response and containment.

- **Predictive maintenance**

  AI systems monitor machinery in real time using sensors and historical data to predict when a machine is likely to fail. This helps in scheduling timely maintenance, reducing downtime, and extending equipment lifespan.

- **Safety and risk management**

  AI-powered surveillance and sensor systems detect unsafe conditions, equipment malfunctions, and worker fatigue, helping to prevent accidents and improve workplace safety.

4.4 Tools and Techniques

In research studies, a wide range of AI and ML tools and techniques are employed to enhance data analysis, prediction, and decision-making processes. Common techniques include supervised learning methods such as regression, decision trees, and support vector machines (SVMs) for classification and prediction tasks, while unsupervised learning methods like clustering and principal component analysis (PCA) help uncover hidden patterns and reduce data dimensionality. Deep learning approaches, including convolutional neural networks (CNNs) and recurrent neural networks (RNNs), are widely used for image analysis, signal processing, and time-series data interpretation. Natural language processing (NLP) tools enable automated extraction and analysis of information from scientific literature and textual datasets. In addition, model evaluation techniques like cross-validation and performance metrics such as accuracy, precision, and F1-score ensure the reliability of results. These techniques are supported by powerful programming libraries and platforms such as TensorFlow, PyTorch, Scikit-learn, and Keras, which streamline the development and deployment of models. These tools and techniques empower researchers to handle complex datasets, derive actionable insights, and accelerate the discovery process across scientific disciplines. In this section, it is discussed how various tools and techniques of AI and ML are employed to enhance data analysis, model complex systems, and derive meaningful insights.

4.4.1 Platforms and Libraries

In research studies, various platforms and libraries support the implementation of AI and ML techniques, enabling efficient data processing, model development, and analysis. Popular platforms used in AI and ML include Google Colab and Jupyter Notebooks are browser-based environments for writing and executing code, with built-in support for Python and ML libraries. Furthermore, cloud-based platforms, such as Amazon SageMaker, Microsoft Azure ML, and IBM Watson Studio, offer scalable infrastructure, automated model training, deployment tools, and integration with large datasets. In addition, RapidMiner and KNIME are drag-and-drop platforms for building ML workflows without extensive programming, often used in data science and business analytics. Furthermore, what libraries are available and where the data can be found are important in AL and ML applications. TensorFlow and Keras are libraries that are widely used for building and training deep learning models, offering flexibility and scalability for both beginners and advanced users. PyTorch is also a library that is preferred for academic research due to its dynamic computation graph and ease of use for complex model development. Other than these, Scikit-learn is also a comprehensive library for classical ML algorithms, including classification, regression, clustering, and model evaluation. Pandas and NumPy are essential for data manipulation and numerical operations, providing the foundation for data preprocessing and analysis. Matplotlib and Seaborn are used for visualizing data distributions, model performance, and insights. These platforms and libraries significantly streamline AI/ML workflows in research, making complex model development more accessible, reproducible, and scalable across disciplines.

4.4.2 Techniques and Algorithms

A variety of AI and ML techniques and algorithms are utilized to analyze complex datasets, make predictions, and uncover hidden patterns. These algorithms enable researchers to build intelligent systems capable of learning from data, improving performance over time, and generating insights that would be difficult to obtain through traditional methods. The most possible used techniques and algorithms in AI and ML are discussed in this section suitable explanation.

- **Machine learning algorithms**
  Supervised learning: Algorithms like linear regression, SVMs, decision trees, and neural networks are used for prediction and classification based on labeled data.

Unsupervised learning: Techniques such as K-means clustering, Hierarchical clustering, and PCA help in pattern recognition and data dimensionality reduction.

Reinforcement learning: This algorithm is applied in optimization problems where agents learn through interactions with environments.

- **Deep learning**

  Convolutional neural networks: CNNs are widely used in image and signal processing tasks such as medical imaging and material characterization.

  Recurrent neural networks and long short-term memory: These learning techniques are useful in time-series analysis, NLP, and forecasting.

- **Natural language processing**

  Techniques such as text mining, sentiment analysis, and entity recognition are NLP techniques and are applied in literature reviews, automated data extraction, and qualitative data analysis.

- **Data preprocessing and feature engineering**

  Tools used for data cleaning, normalization, transformation, and selection of relevant features are Scikit-learn, Pandas, and NumPy.

- **Model evaluation and validation**

  Metrics like accuracy, precision, recall, F1-score, and cross-validation techniques ensure robust model performance, evaluation, validation, and generalization.

- **Automation and optimization**

  Techniques like genetic algorithms (GAs), swarm intelligence, and Bayesian optimization are used for design, simulation, and hyper-parameter tuning.

These tools and techniques collectively empower researchers to handle large datasets, uncover hidden patterns, automate routine tasks, and develop predictive and prescriptive models to support evidence-based conclusions. Here's how AI and ML tools and techniques are applied in a specific research field.

4.5 Advantages of AI and ML

Pros: Some of the advantages of AI and ML are as follows:

1. Human error could be minimized and eliminated.
2. It could be available 24 × 7. There is no issue of tiredness/relaxation.

3. Could be used efficiently for repetitive tasks, and there is not much creativity needed.

4. Generally, it does work fast compared to humans, especially in the case of large datasets.

5. It is capable of generating many combinations of process parameters and suggests the best one for real-life applications based on the datasets used.

Cons: Some of the disadvantages of AI and ML are as follows:

1. Relatively costly and skilled workers are needed to do the programming and monitoring of the working of the models.

2. Could not replace humans fully due to less creativity; limited to repeated tasks.

3. It could replace some jobs, which might lead to unemployment.

4. This cannot identify the mistakes own but needs a skilled worker to analyze the outcomes of these models.

4.6 Ethical Consideration

The application of AI and ML in research studies brings with it several ethical considerations that must be carefully addressed to ensure responsible and unbiased use. Data privacy and confidentiality are primary concerns, as AI/ML systems often rely on large datasets that may include sensitive personal or proprietary information. Researchers must ensure compliance with data protection regulations such as the General Data Protection Regulation (GDPR) or Health Insurance Portability and Accountability Act (HIPAA) and implement secure data handling practices. Bias and fairness also pose significant challenges; algorithms trained on unrepresentative or biased data can lead to skewed results and reinforce existing inequalities, especially in healthcare, social sciences, and criminal justice research. Transparency is another ethical imperative in AI and ML models, particularly deep learning systems, which often function as "black boxes," making it difficult to interpret how decisions are made. This lack of explainability can undermine trust in research findings. Furthermore, there is a need for accountability in the deployment and impact of AI models; researchers must be clear about the limitations and potential consequences of their systems. Lastly, the use of AI should not replace human judgment but rather augment it, ensuring that scientific rigor and ethical responsibility are maintained throughout the research process. The following important factors should be practiced during the application of AI and ML in research.

Data privacy and confidentiality
- Ensure secure handling of sensitive data.
- Comply with legal frameworks.

Bias and fairness
- Address potential biases in datasets and algorithms.
- Avoid reinforcing social, gender, or racial inequalities.

Transparency and explainability
- Strive for model interpretability, especially in critical decision-making.
- Communicate how AI/ML models generate results.

Accountability
- Clearly define responsibility for AI-generated outcomes.
- Acknowledge limitations and uncertainties in models.

Informed consent
- Obtain proper consent for using data, especially in human subject research.
- Ensure participants are aware of how AI/ML will be used.

Avoiding over-reliance on automation
- Use AI/ML to support, not replace, scientific and ethical judgment.
- Retain human oversight in key decisions.

Impact on society and research integrity
- Consider long-term societal consequences of AI applications.
- Maintain transparency and integrity in research dissemination.

4.7 Future of AI and ML in Research

The future of AI and ML in research is to be transformative, pushing the boundaries of discovery across disciplines, and their synergy with human creativity is leading to a new era of innovation, discovery, and scientific advancement. The integration of AI and ML will also promote more collaborative and interdisciplinary research effectively and within a specific timeframe. Ethical use of AI will become a critical focus, ensuring that algorithms used in sensitive fields like healthcare, social science, and policy research are transparent, unbiased, and accountable. As these technologies become more sophisticated, they will not only automate routine tasks but also enable deeper insights through predictive modeling, real-time data analysis, and autonomous experimentation.

The AI algorithms could be more advanced and accessible in the near future; they will play a crucial role not just in analyzing data but providing new insight, such as automating hypothesis generation, suggesting novel research directions, and even designing experiments. This will accelerate the pace of discovery in fields like medicine, environmental science, physics, and social sciences. Furthermore, the usage of AI with emerging technologies like quantum computing, IoT, and synthetic biology will open new frontiers for interdisciplinary research. In education and academia, AI will support personalized learning and mentoring for researchers, helping them develop skills tailored to their specific areas of study. However, this future also demands careful attention to ethics, transparency, and data governance, ensuring that AI-driven research remains trustworthy, equitable, and inclusive. As AI continues to evolve, it will not only enhance human intellect but also redefine what's possible in the pursuit of knowledge. The scope of AI and ML could be endless if it is used properly and ethically for the benefit of society.

Theoretical Questions and Problems

1. How could ML techniques be used for data processing? Explain the process in detail with advantages and limitations.
2. How could AI techniques be used for data processing? Explain the process in detail with advantages and limitations.
3. What is the difference between ML and AI? Which one is best for data processing and why?
4. What could be the future of ML and AI in the context of the research?
5. What is the difference between ML, deep learning, and AI? Explain briefly with a neat sketch.
6. Explain the procedure/steps for identifying the ethical use of AI and ML in the research.

7. Write the applications of AI and ML in the different fields of Science and Engineering.

8. Differentiate between AI, ML, and deep learning. How do we select them for our research?

9. What could be the future of AI, ML in the field of research?

5

Fabrication and Manufacturing Techniques

5.1 Introduction

Manufacturing is the foundation of human life, as the survival of human beings is impossible without it. Fabrication and manufacturing techniques are not new; they have an extensive history similar to mankind in this world. The foundry/casting is one of the oldest manufacturing processes; it started in the Mesopotamian period. Furthermore, other manufacturing techniques were developed and play vital roles in modern life and enable the creation of everything from necessities like food and shelter to advanced technologies that drive innovation and improve the quality of life. Advancement in manufacturing techniques depends on the requirements of the consumer. Important fabrication and manufacturing techniques are to be understood with their processing mechanism, benefits, limitations, and applications before conducting any research work.

5.2 Fabrication Techniques

Fabrication techniques are also manufacturing processes, but they are limited to producing the final product by assembling the semi-finished components and shaping them into the final product. The type of raw materials and product applications makes them different from general manufacturing processes. Fabrication techniques other than manufacturing (like- cutting, forming, joining, etc.) are as follows.

5.2.1 Physical Vapor Deposition

Physical Vapor Deposition (PVD) techniques were developed in 1995 to improve the corrosion and wear-resistant properties of the base materials for various industrial applications. The surface behavior of the metallic materials can be modified by the deposition of hard materials layers through PVD. As

DOI: 10.1201/9781003664871-5

a convincing outcome of PVD in corrosion, wear, and fatigue improvements, this process has started to be implemented in the biomedical field to enhance bioactivity and bone regeneration. Physical vapor deposition is a versatile vacuum deposition method used to produce thin films and coatings on various substrates, including metals, ceramics, glass, and polymers. PVD, warm, or hot vacuum techniques are very extensive and diverse, able to produce coatings through high-temperature reactions. However, material scientists need to pay more attention to the selection of substrate materials compatible with the process. Figure 5.1 shows the schematic of the PVD process, which comprises a substrate holder, target boat, sputtering gas (usually inert gas), vacuum system, and power supply unit. The target material vaporizes and is deposited over the substrate surface with the help of the vaporization process.

5.2.1.1 Working Principles

The PVD technique typically involves three fundamental steps for thin-film coating. The transition of the target material from a condensed phase to a vapor phase, transport of vapor on the substrate surface, and then back to a

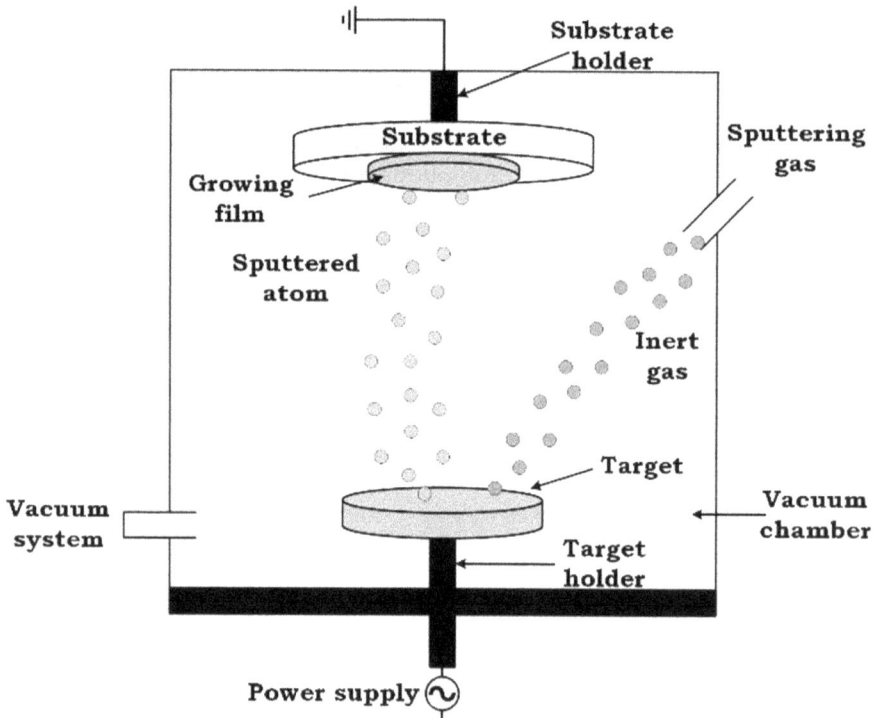

FIGURE 5.1
Schematic of physical vapor deposition of target material on the substrate material.

condensed phase as a thin film on the substrate material. The entire process must be conducted in a vacuum environment. The primary steps are vaporization of source materials, transport of vaporized atoms, and deposition on substrate.

5.2.1.2 Types of Physical Vapor Deposition Techniques

The PVD technique is classified as per the source of energy used to vaporize the solid material, and these are as follows:

1. **Thermal evaporation**: The target material is heated until it evaporates, and then, the vapor is deposited on the substrate.
2. **Sputtering**: Energetic ions bombard the target material, which ejects atoms from the target and deposits them on the substrate.
3. **Electron beam evaporation**: A high-energy electron beam heats the material, causing it to evaporate and deposit on the substrate.
4. **Pulsed laser deposition**: A high-power laser ablates material from the target, creating a vapor that condenses on the substrate.
5. **Close-space sublimation**: The material and substrate are placed close together and heated radiatively.

5.2.1.3 Applications of Physical Vapor Deposition Technique

PVD is widely used in producing thin films/coatings with specific characteristics for various industrial applications. This technique is one of the best fabrication techniques in the

- **Electronic industry**: Used to fabricate semiconductor devices, thin-film solar cells, micro-electro-mechanical systems (MEMS), and other electronic components.
- **Optical coating industry**: Used this technique for anti-reflective coatings on lenses and mirrors, sustainable building glass coatings, and other customized coatings.
- **Cutting tools manufacturing**: Coating of ceramic/cermets on cutting tools to enhance hardness and durability.
- **Packaging**: Helpful in the packaging industry to make aluminized PET film for food packaging and balloons.

5.2.2 Chemical Vapor Deposition

Chemical vapor deposition (CVD) is a widely used technique for producing high-quality, high-performance solid materials, particularly for thin-film deposition. CVD is a fundamental tool in modern manufacturing, enabling the

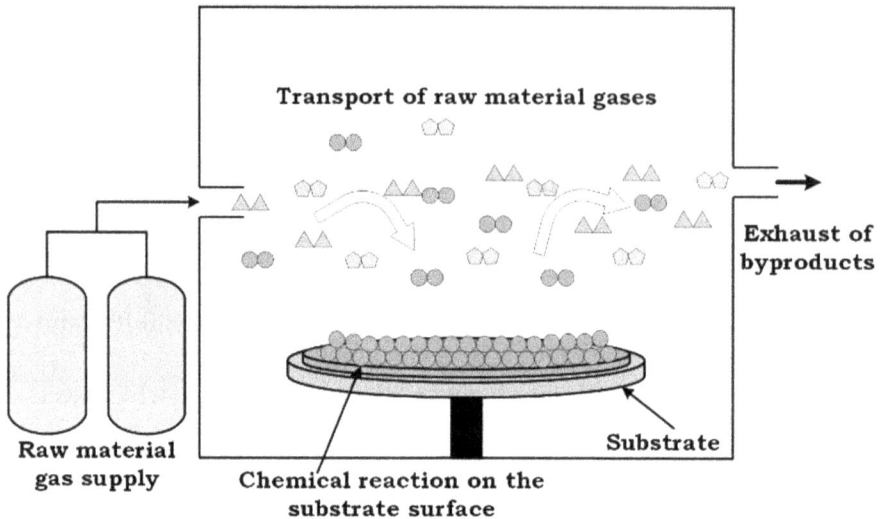

FIGURE 5.2
Schematic representation of the CVD process.

fabrication of advanced materials and devices with tailored properties. This technique involves the chemical reactions of gaseous precursors on a heated substrate to form a solid material. Figure 5.2 shows the schematic of the CVD process used for enhancing the surface properties of industrial products.

5.2.2.1 Working Principles

The fundamental requirements of this process are reacting gases required to deposit over the substrate surface, adsorption of gases by the substrate and chemical reactions on the substrate surface, and desorption of by-products of the chemical reaction through gas flow. Details of the working principle of the CVD process are explained, and the CVD technique is classified into five categories according to their process physics.

- **Transport of reacting gases**: Gaseous precursors are introduced into a reaction chamber as target materials for thin-film deposition.
- **Adsorption**: In this process, the precursor gases are adsorbed onto the heated substrate surface.
- **Surface reactions**: After adsorption, chemical reactions occur on the substrate surface, leading to the formation of a solid film on the substrate.
- **Desorption**: This technique depends on chemical reactions between precursor gases and substrate; hence, by-products of the reaction are needed to be desorbed and removed from the chamber by gas flow.

5.2.2.2 Types of Chemical Vapor Deposition Technique

1. **Thermally activated CVD**: Uses heat to initiate chemical reactions. The gaseous precursors are thermally decomposed on the heated substrate.
2. **Plasma-enhanced CVD**: Uses plasma to enhance chemical reactions at lower temperatures. This is useful for depositing materials on temperature-sensitive substrates.
3. **Low-pressure CVD**: Conducted at sub-atmospheric pressures to improve film uniformity and reduce unwanted gas-phase reactions.
4. **Atmospheric pressure CVD**: Conducted at atmospheric pressure, often used for large-scale production.
5. **Ultrahigh vacuum CVD**: Conducted at very low pressures, typically below 10^{-6} Pa, to achieve high-purity films.

5.2.2.3 Applications of Chemical Vapor Deposition

CVD is used in various industries for producing thin films with specific properties. Some of these include semiconductor industries, such as the deposition of silicon dioxide, silicon nitride, and other materials for integrated circuits. Producing hard coatings on cutting tools and wear-resistant coatings on various surfaces. Depositing anti-reflective coatings on lenses and mirrors in optical applications. In nanomaterials development, CVD is used in growing carbon nanotubes, graphene, and other nanomaterials for various modern applications.

5.2.2.4 Advantages and Challenges of Chemical Vapor Deposition

- **High purity**: It produces high-purity thin films with excellent uniformity.
- **Versatility**: It can be used with a wide range of materials and substrates.
- **Scalability**: suitable for both small-scale research and large-scale industrial production.
- **Complexity**: It requires precise control of process parameters, sophisticated equipment, and skilled manpower.
- **Environment**: Some CVD processes involve hazardous chemicals that require careful handling and disposal.

5.2.3 Lithography

Lithography is a versatile printing technique that involves creating specific features on a flat surface. This technique was invented by German playwright and actor Alois Senefelder in the 1800s. Lithography refers to the fabrication of 1D and 2D structures in which at least one of the lateral dimensions is

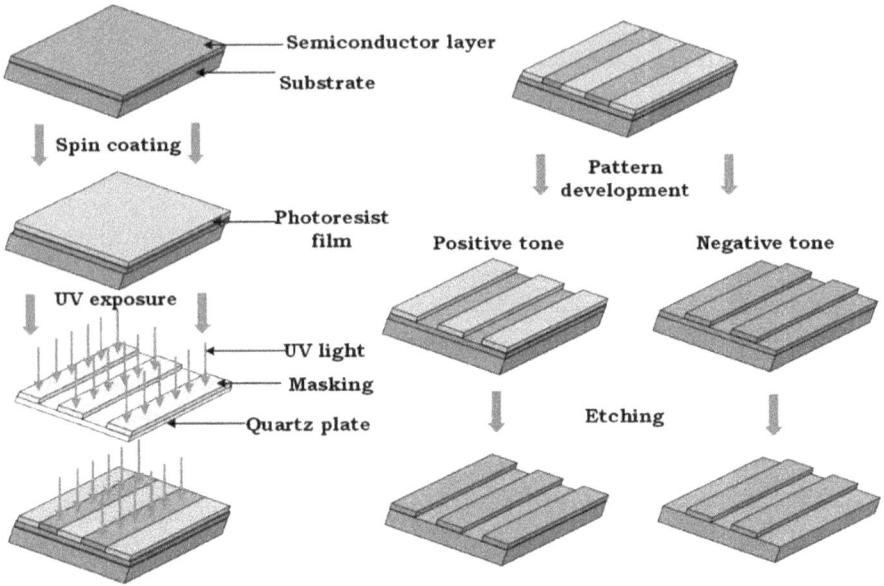

FIGURE 5.3
Steps of the photolithography process: substrate preparation, photoresist coating, masking and exposure to UV radiation, pattern development and stripping, chemical treatment, and final structure development.

on the nanometer scale. Lithography replicates the required patterns (can be positive and negative masks) into the substrate. Lithography became a popular fundamental technique for both art and industry, enabling the creation of detailed and high-quality prints across various applications. Figure 5.3 shows the working of the photolithography process.

5.2.3.1 Working Principle of Lithography

The primary principle of lithography is to transfer a pattern of any geometric shapes in a mask to thin layer of radiation sensitive material. There are the following steps are to be followed:

- **Preparation of substrate**: Substrate preparation is generally required in semiconductor applications as the first step.
- **Photoresist coating**: After substrate preparation, photoresist coating is to be done to make the pattern.
- **Masking**: It is a pattern that is to be printed into the underlying substrate. It can be a positive and a negative mask.
- **Chemical treatment**: After masking, the surface is treated with a chemical solution (etchant) to fix the pattern in place.

5.2.3.2 Types of Lithography

1. Photolithography
2. Electron beam lithography
3. X-ray and UV lithography
4. Focused ion beam lithography
5. Soft lithography
6. Nanoimprint lithography

5.2.3.3 Applications of Lithography

Lithography is used in producing detailed and intricate prints for artists, printing books, magazines, posters, and advertisements. It is widely used in microelectronics to create patterns on semiconductor wafers for integrated circuits. Photolithography and soft lithography are often applied to construct scaffolds for neural tissue engineering.

5.2.4 Sputtering

Sputtering is a versatile and widely used physical deposition technique for depositing thin films of target materials onto substrates. The fundamental working principle of sputtering involves bombarding a target material (cathode) with energetic ions, typically from a plasma, causing atoms from the target to be ejected and deposited onto a substrate. This process can be broken down into several key steps:

- **Plasma creation**: In this step, plasma is created by ionizing a gas, usually argon, using an electric field or electromagnetic waves. The plasma consists of positively charged ions and electrons.
- **Ion acceleration**: Ions present in the plasma are accelerated toward the target material through an electric field, possessing kinetic energy; these ions travel toward the target.
- **Target bombardment**: High-energy ions collide with the target material, transferring their energy to the target atoms. This collision of ions and atoms causes the ejection of atoms from the target surface.
- **Film deposition**: Ejected atoms from the target surface travel through the vacuum and condense onto the substrate and form a thin film.

5.2.4.1 Types of Sputtering

1. **Direct current sputtering**: It uses a direct current (DC) to create the plasma and is suitable for conductive materials.

2. **Radio frequency sputtering**: Radio frequency is the source to create the plasma and is suitable for insulating materials.

3. **Magnetron sputtering**: A magnetic field is used to trap electrons near the target surface, increasing ionization efficiency and deposition rates.

4. **Ion beam sputtering**: An ion beam helps in the ejection of atoms from the target to achieve precise control over the sputtering process.

5.2.4.2 Applications of Sputtering

Sputtering is used in various industries for producing thin films with specific properties. Some applications include semiconductor manufacturing, optical coatings, thin-film solar cells, and decorative coatings for aesthetic purposes on various surfaces.

5.2.5 Coating

Coating refers to the application of a layer of material onto a surface to enhance its properties, such as protection against corrosion, wear resistance, environmental damage, and aesthetic appeal. Coating processes are divided into the following categories. These processes are critical in various industries, ensuring that products are not only functional but also durable and aesthetically pleasing.

1. **Powder coating**

   Powder coating involves applying electrostatically charged powder particles to a surface, which are then cured under heat to form a hard, durable finish. This coating technique is used for automotive parts, appliances, and outdoor furniture. Powder coating provides a high-quality finish and excellent durability.

2. **Spray coating**

   Liquid coating is sprayed onto a surface using a spray gun, creating an even layer. This is used in automotive refinishing, industrial coatings, and decorative finishes. It is versatile, easy to apply, and suitable for a wide range of materials.

3. **Dip coating**

   In this coating technique, the substrate is immersed in a liquid coating solution and then withdrawn, allowing the coating to dry and form a layer. This process is useful for applying protective coatings to small parts, electronics, and ceramics as it is a simple process and provides uniform coating and good adhesion.

4. **Electroplating**

   A metal coating is deposited onto a substrate by passing an electric current through an electrolytic solution containing metal ions.

It is used for enhancing corrosion resistance, improving appearance, and adding wear resistance to metal parts. Electroplating provides a strong, uniform coating and can be used with various metals.

5. **Electrostatic spray coating**

   Similar to spray coating, the coating material is charged electrostatically to improve adhesion and reduce overspray. It is very useful for automotive refinishing, industrial coatings, and consumer goods with improved efficiency, reduced waste, and better coverage.

5.2.5.1 Applications of Coating

Coating processes are used across a wide range of industries, from automotive to electronics. This process can add specific properties, such as electrical conductivity or an insulating layer on the base material. Hard material coating extends the lifespan of the underlying material by providing a robust protective layer.

5.3 Manufacturing Techniques

Manufacturing is a broader term that encompasses the entire process of producing goods, from raw materials to finished products. It not only includes fabrication but also involves other processes such as casting, molding, welding, subtractive manufacturing, and additive manufacturing, and mass production assembly lines. Manufacturing techniques play a vital role in producing consumer products such as electronics, automobiles, household appliances, and packaged foods, and also in setting up industrial plants. There are the following advanced manufacturing processes, widely used in specific and precision work, customized products, and research and development work.

5.3.1 Electrical Discharge Machining

Electrical discharge machining (EDM) is a type of subtractive manufacturing process, which is a well-stabilized technique in the machining of metals. It is a non-traditional machining process that uses electrical discharges to erode material from a workpiece. Hence, this process is also known as spark machining or spark erosion. EDM works by creating a series of rapid electrical discharges (sparks) between an electrode (tool) and the workpiece, which are both submerged in a dielectric fluid. The sparks generate intense heat, which melts and vaporizes the material from the workpiece. The shape of the machined workpiece is similar to the tool shape. This technique is used for machining complex shapes and hard materials that are difficult to process

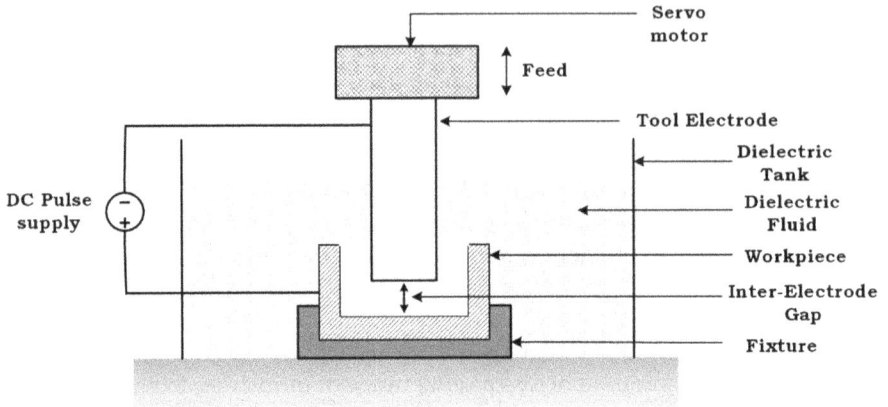

FIGURE 5.4
Schematic diagram of the die-sinking electron discharge machining setup.

using traditional methods. The schematic diagram of the die-sinking EDM setup is shown in Figure 5.4.

5.3.1.1 Working Principle

- **Machining setup**: The workpiece and the electrode are submerged in a dielectric fluid, which is generally deionized water or oil.
- **Gap maintenance**: A small gap is maintained between the electrode and the workpiece for generating sparks.
- **Pulse discharge**: A power supply generates a series of rapid electrical pulses, creating sparks between the electrode and the workpiece.
- **Material removal**: The sparks generate intense heat, melting and vaporizing the material from the workpiece.
- **Cooling**: The dielectric fluid cools and flushes away the eroded material, maintaining the gap and preventing short circuits.

5.3.1.2 Types of Electrical Discharge Machining

1. **Wire electrical discharge machining**: It is used to cut intricate shapes and patterns in the workpiece using a thin wire as the electrode.
2. **Hole drilling electrical discharge machining**: This EDM is specialized for drilling small holes, precise holes in hard materials.
3. **Die sinking EDM**: It is used for making cavities in the workpiece using a die tool (as an electrode) of a similar shape and size.

5.3.1.3 Applications of Electrical Discharge Machining

- **Aerospace**: Machining complex components and molds.
- **Automotive**: Producing precision parts and molds.
- **Medical**: Creating intricate medical devices and implants.
- **Toolmaking**: Fabricating molds, dies, and other precision tools.

5.3.1.4 Advantages of Electrical Discharge Machining

- **Precision**: EDM is capable of achieving high accuracy and fine details.
- **Non-contact**: There is no mechanical stress on the workpiece, which reduces wear and tear.
- **Versatility**: It is suitable for hard and heat-resistant materials.
- **Complex shapes**: This process can create intricate geometries and sharp internal corners.

5.3.1.5 Challenges of Electrical Discharge Machining

- **Slow process**: EDM is generally slower than conventional machining methods and needs more machining time.
- **High energy consumption**: Due to the longer machining time, this process requires significant electrical power.
- **Tool wear**: In this process, the tool acts like an electrode, and due to spark generation, wear occurs on the electrode and needs regular replacement.

5.3.2 Electrochemical Machining

Electrochemical machining (ECM) is a non-conventional machining process that uses electrochemical reactions to remove material from a workpiece. ECM involves the controlled dissolution of material from a workpiece by applying an electric current between the tool (cathode) and the workpiece (anode) submerged in an electrolyte solution. The process is often referred to as "reverse electroplating" because it removes material instead of adding it. ECM is a powerful technique for machining complex shapes and hard materials that are difficult to process using traditional methods. Figure 5.5 shows the sketch of the ECM setup.

5.3.2.1 Working Principle

- **Machining setup**: The workpiece and the tool are submerged in an electrolyte solution, typically a conductive fluid like sodium chloride or sodium nitrate.

- **Electrolyte flow**: The electrolyte is pumped through the gap between the tool and the workpiece at a controlled rate.
- **Electric current**: A DC is applied between the tool and the workpiece. The workpiece acts as the anode, and the tool acts as the cathode.
- **Material removal**: As the electric current flows, metal ions from the workpiece dissolve into the electrolyte and are carried away, leaving behind the desired shape.
- **Gap maintenance**: The gap between the tool and the workpiece is maintained by controlling the feed rate of the tool and the flow of the electrolyte.

5.3.2.2 Classification of Electrochemical Machining

1. **Electrochemical grinding**: This process uses a grinding wheel with an insulating abrasive, like diamond particles, as a cutting tool, and the rest of the process is similar to regular ECM.
2. **Electrochemical polishing**: This is also called electrolytic polishing; this process improves the surface finish of metal parts by dissolving the anode.
3. **Electrochemical jet machining**: It is also a variant of ECM and uses a jet of electrolyte to dissolve material from a workpiece.

5.3.2.3 Applications of Electrochemical Machining

- **Die-sinking**: ECM is widely used in the manufacturing of cavities in the workpiece.
- **Drilling**: ECM can drill multiple holes, including jet engine turbine blades.
- **Machining**: ECM can machine steam turbine blades and other parts.
- **Profiling and contouring**: ECM can be used for profiling and contouring.
- **Rifling barrel**: ECM can be used for rifling barrels.

5.3.2.4 Advantages of Electrochemical Machining

- **Complex shapes**: It can machine complex two-dimensional shapes.
- **Independent of material properties**: It is independent of the electrical or thermal properties of the material being machined than EDM.
- **High surface finish**: It can produce a high surface finish because dissolution occurs at the atomic level.
- **No tool wear**: ECM has little or no tool wear than EDM.
- **Remove deposits**: It can remove deposits like oxides and coatings from surfaces.

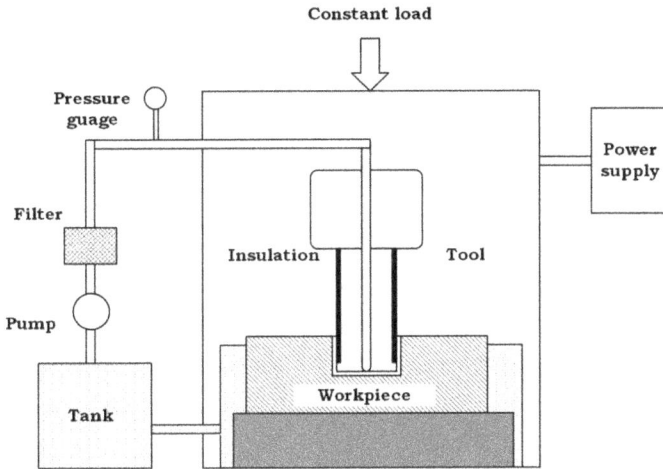

FIGURE 5.5
Schematic diagram of ECM setup.

5.3.2.5 Challenges of Electrochemical Machining

- **Initial tooling cost**: High initial tooling cost for ECM and time-consuming.
- **Not suitable for soft materials**: It is not suitable for soft or ductile materials.
- **High power consumption**: Requires a lot of power to drive the grinding wheel and pump.
- **Environmental risk**: It can produce environmentally harmful by-products.

5.3.3 Electron Beam Machining

Electron beam machining (EBM) is a non-conventional machining process that uses a high-velocity electron beam to remove material from a workpiece. EBM involves focusing a beam of high-velocity electrons onto a small area of the workpiece. The kinetic energy of the electrons is converted into heat, which vaporizes the material from the workpiece. The working principle of the EBM process is explained as follows.

5.3.3.1 Working Principle

- **Electron generation unit**: An electron gun is used to generate high-velocity electrons by heating a tungsten filament (cathode) and accelerating them using a high-voltage power supply.

- **Electromagnetic lenses**: It is used for the precise machining of the workpiece; the electrons need to be focused into a narrow beam. Electromagnetic lenses are used to make the desired size of the electron beam.
- **Deflectors**: These are used to direct the electron beam to the desired location on the workpiece.
- **Material removal**: In EBM, the focused electron beam creates intense heat, which vaporizes the material from the workpiece.
- **Vacuum environment**: The entire process is carried out in a vacuum chamber to prevent electron collisions with air molecules and to minimize contamination.

5.3.3.2 Process Parameters

- **Beam current**: Number of electrons in the beam, typically ranging from 200 μA to 1 A.
- **Pulse duration**: Duration of each pulse, which can range from 50 μs to 15 ms.
- **Energy density**: Energy per pulse, which can exceed 100 J/pulse.
- **Spot size**: Diameter of the focused beam, typically between 0.01 and 0.02 mm.

5.3.3.3 Applications of Electron Beam Machining

- **Aerospace**: EBM is used to create lightweight yet strong components for aircraft and spacecraft, such as turbine blades for jet engines.
- **Medical**: EBM is used to make medical implants and prosthetics, such as hip cups, acetabular cups, and hemipelvic implants. It is well-suited to this industry and can produce biocompatible parts with precise manufacturing.
- **Tooling and prototyping**: EBM can reduce lead times and speed up development cycles.
- **Other industries**: EBM is also used in the automotive, defense, microelectronics, nuclear, jewelry, and watchmaking industries.

5.3.3.4 Advantages of Electron Beam Machining

- **High quality**: It produces high-quality metal parts with an excellent strength-to-weight ratio.
- **Accuracy**: It has excellent dimensional accuracy and surface finish.
- **Complex geometries**: It produces complex and intricate shapes.

- **Minimal heat-affected zones**: It preserves material integrity with minimal heat-affected zones.
- **No tool wear**: It reduces tooling costs by eliminating mechanical tool wear.
- **Environmental impact**: It has minimal environmental impact because it doesn't use coolants or lubricants.

5.3.3.5 Challenges of Electron Beam Machining

- **High equipment cost**: EBM has a high equipment cost.
- **Limited material selection**: It has a limited material selection.
- **Slower build speed**: EBM has a slower build speed.
- **Hazardous X-rays**: EBM produces hazardous X-rays.
- **Vacuum control**: It is very difficult to maintain a perfect vacuum in EBM.
- **Suitability**: This is only suitable for small workpieces.
- **Skill**: It needs trained professionals to operate EBM.

5.3.4 Laser Beam Machining

Laser beam machining (LBM) is a non-conventional machining process that uses a high-intensity laser beam for material removal from a workpiece. The energy from the laser beam is absorbed by the material, causing it to heat up, melt, and vaporize, thereby removing material from the workpiece. Figure 5.6 depicts the working principle of LBM and the mechanism of laser and material interaction.

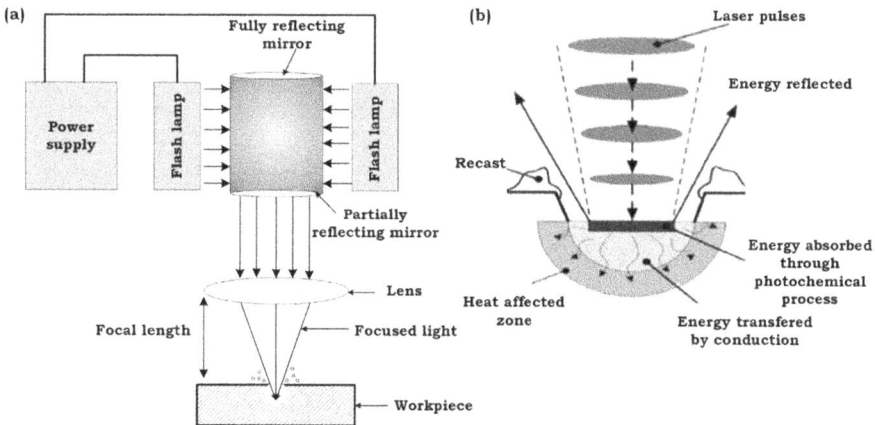

FIGURE 5.6
Schematic diagram of (a) laser beam machining and (b) mechanism of laser and material interaction.

5.3.4.1 Working Principle

- **Laser generation**: A laser source, such as CO_2, Nd: YAG, or fiber laser, generates a coherent and collimated beam of light.
- **Beam focusing**: The laser beam is focused using optical lenses to a small spot on the workpiece, concentrating the energy.
- **Material interaction**: When the laser beam strikes the workpiece, the material absorbs the energy, leading to rapid heating and localized melting.
- **Material removal**: The intense heat causes the material to vaporize or melt, and the molten material is ejected from the cutting zone by an assist gas, such as oxygen or nitrogen.

5.3.4.2 Process Parameters

- **Laser power**: Determines the intensity of the laser beam and affects the material removal rate.
- **Focus spot size**: The size of the focused laser spot influences the precision and quality of the cut.
- **Assist gas**: Types of assist gases (e.g., oxygen, nitrogen) used to assist in material removal and cooling.
- **Cutting speed**: The speed at which the laser moves across the workpiece, affecting the quality and efficiency of the process.

5.3.4.3 Types of Laser Beam Machining

1. **Laser cutting**: Uses a laser to cut through materials, suitable for metals, ceramics, plastics, and composites.
2. **Laser drilling**: Uses a laser to create small, precise holes in materials.
3. **Laser engraving/marking**: Uses the laser to remove material from the surface to create patterns or text.

5.3.4.4 Applications of Laser Beam Machining

- **Aerospace**: LBM is used for cutting and drilling of high-strength alloys and composite materials. Also makes jet engine components, titanium ducting, and fuselage skin stiffeners.
- **Automotive**: It is used to make airbags and body frames.
- **Medical**: It is used to make microfluidic devices, coronary stents, and surgical equipment.
- **Electronics**: LBM is used for circuiting, line stripping, precision cutting, and structuring of printed circuit boards (PCBs) and semiconductor wafers.

5.3.4.5 Advantages of Laser Beam Machining

- **Accuracy**: LBM is very precise and accurate.
- **Material range**: It can work with a wide range of materials, including metals, ceramics, and polymers.
- **Non-contact**: It is a non-contact machining process.
- **Minimal waste**: LBM produces minimal material waste.
- **Small holes**: LBM is often used to machine small-diameter holes because the laser beam can be focused within a small range.

5.3.4.6 Challenges of Laser Beam Machining

- **Maintenance**: LBM requires a lot of money to maintain.
- **Thickness**: LBM has limitations on material thickness due to taper.
- **Cost**: LBM is expensive compared to other cutting techniques.
- **Gases and fumes**: LBM can produce harmful gases and fumes.
- **Energy consumption**: LBM consumes a lot of energy.
- **Suitability**: Not all metals can be cut with a laser machine, such as copper and aluminum.
- **Burns**: Improper settings on the laser machine can cause burns.

5.3.5 Plasma Arc Machining

Plasma arc machining (PAM) is a non-conventional machining process that uses a high-velocity jet of ionized gas (plasma) to remove material from a workpiece. PAM involves creating a plasma arc by passing a gas through an electric arc between a cathode and an anode. The plasma arc generates temperatures ranging from 11,000°C to 30,000°C, which is sufficient to melt and vaporize the material from the workpiece.

5.3.5.1 Working Principles

- **Setup design**: The workpiece and the plasma torch are positioned appropriately.
- **Gas supply**: A gas (such as nitrogen, argon, hydrogen, or a mixture) is supplied to the plasma torch.
- **Arc generation**: A high-frequency spark is created between a tungsten electrode (cathode) and a copper nozzle (anode), both of which are water-cooled.
- **Ionization**: The high-velocity electrons generated by the arc collide with the gas molecules, causing ionization and producing a high-temperature plasma jet.

- **Material removal**: The plasma jet is directed onto the workpiece, melting and vaporizing the material. The molten material is then blown away by the high-velocity gas stream.
- **Cooling**: The plasma torch and the workpiece are cooled using a cooling water system to prevent overheating.

5.3.5.2 Process Parameters

- **Gas type**: The type of gas used affects the plasma characteristics and the quality of the cut.
- **Current**: The electrical current determines the intensity of the plasma arc and the material removal rate.
- **Nozzle design**: The design of the nozzle affects the focus and stability of the plasma jet.
- **Standoff distance**: The distance between the nozzle and the workpiece influences the precision and quality of the cut.

5.3.5.3 Applications of Plasma Arc Machining

- **Cutting**: It can cut a variety of materials, including stainless steel, aluminum, titanium, copper, and nickel, and can also be used for underwater cutting.
- **Welding**: PAM can be used to weld materials like stainless steel and titanium.
- **Aerospace and defense**: PAM is used in the aerospace and defense industries for applications like welding steel rocket motor cases and working with nuclear submarine pipe systems.
- **Medical device manufacturing**: This process is used in the manufacturing of medical devices, especially for intricate components.
- **Automotive**: PAM is used in the automotive industry to fabricate critical components like engine parts and exhaust systems.
- **Power generation**: This is used in power generation equipment because it can work with high-temperature materials.

5.3.5.4 Advantages of Plasma Arc Machining

- **Versatility**: PAM can be used on a variety of materials, including aluminum, stainless steel, copper, and brass.
- **Precision**: PAM produces precise cuts and accurate dimensions.
- **Cutting capacity**: The more energy added to the plasma arc, the hotter it becomes, which increases cutting capacity.
- **Faster speeds**: PAM can cut multiple materials stacked on top of each other at high speeds.

5.3.5.5 Challenges of Plasma Arc Machining

- **Fumes**: PAM can produce dangerous fumes that need to be controlled.
- **Noise**: PAM can be noisy.
- **Shallow cuts**: PAM can only cut to a certain depth, so it may not be able to cut through thick or dense materials.
- **Equipment cost**: It requires significant investment in plasma torch equipment and maintenance.
- **Energy consumption**: High energy consumption compared to some conventional methods.
- **Maintenance**: PAM equipment requires skilled technicians for maintenance.

5.3.6 Abrasive Jet Machining

Abrasive jet machining (AJM) is a non-traditional machining process that uses a high-velocity stream of abrasive particles carried by a gas to remove material from a workpiece. It's particularly useful for cutting, cleaning, and engraving hard and brittle materials. AJM operates by directing a high-velocity jet of abrasive particles mixed with a gas (usually air) toward the workpiece. The impact of the abrasive particles erodes the material from the workpiece surface.

5.3.6.1 Working Principle

- **Gas supply**: A high-pressure gas, often air or nitrogen, is supplied to the system.
- **Mixing chamber**: The high-pressure gas is mixed with fine abrasive particles, such as aluminum oxide (Al_2O_3), silicon carbide (SiC), or glass beads, in a mixing chamber.
- **Nozzle**: The abrasive gas mixture is then directed through a nozzle, focusing the high-velocity stream onto the workpiece.
- **Material removal**: The kinetic energy of the abrasive particles impacts the workpiece surface, which causes micro-cutting and erosion, and further removes material. Brittle materials are generally machined using AJM.

5.3.6.2 Process Parameters

- **Abrasive type**: The type and size of abrasive particles affect the material removal rate and surface finish.
- **Gas pressure**: The pressure of the gas determines the velocity of the abrasive particles and influences the cutting efficiency.

- **Nozzle design**: The design and diameter of the nozzle affect the focus and intensity of the abrasive jet.
- **Stand-off distance**: The distance between the nozzle and the work-piece impacts the accuracy and effectiveness of the process.

5.3.6.3 Applications of Abrasive Jet Machining

- **Cutting and drilling**: Used for cutting and drilling hard and brittle materials like glass, ceramics, and quartz.
- **Cleaning**: Effective for cleaning molds, removing oxide layers, and surface preparation.
- **Deburring**: Removing burrs from machined parts.
- **Engraving**: Used for engraving intricate patterns and designs on hard materials.

5.3.6.4 Advantages of Abrasive Jet Machining

- **Flexibility**: Uses hoses to transport gas and abrasive to any part of the workpiece, including hard-to-reach areas.
- **Minimal thermal damage**: Reduces the risk of thermal damage to the workpieces.
- **Versatility**: Suitable for a wide range of hard and brittle materials.
- **Machinability**: Can be used for drilling, cutting, deburring, deflashing, cleaning, and decorating.
- **Low investment**: Relatively low set-up costs as compared to other processes.

5.3.6.5 Challenges of Abrasive Jet Machining

- **Nozzle wear**: Nozzles wear out quickly and need frequent replacement.
- **Dust generation**: Produces dust and requires proper ventilation and filtration systems.
- **Low material removal rate**: Not suitable for high material removal rates.
- **Produces tapered cuts**: Can cause surface roughness and burrs.
- **Not suitable for soft materials**: Abrasive particles can get embedded in the work surface, which can affect cut quality.

5.3.7 3D Printing/Additive Manufacturing

Additive manufacturing (3D printing, rapid prototyping, and digital manufacturing) is one of the advanced manufacturing techniques that was theorized in the 1970s. It is a transformative technology that builds 3D objects through a layer-by-layer structure using an STL file. Unlike traditional subtractive manufacturing, which removes material from a solid block, additive manufacturing adds material layer by layer to create the final product.

5.3.7.1 Working Principles of Additive Manufacturing

The working principle of this technique is the addition of material in the form of layer-by-layer printed material into the desired product and shape with high accuracy. The required steps of additive manufacturing are explained as follows:

- **Design creation**: Create a digital 3D model, using Computer-Aided Design (CAD) software and any other designing software.
- **STL file conversion and slicing**: The CAD file needs to be converted into an STL file. Afterward, slice the 3D CAD model into thin horizontal layers using specialized software. Each layer will show the cross-section of the object.
- **Setting process parameters**: Set the process parameters as per the printer specifications and upload the final STL file into the 3D printer for printing.
- **Printing**: The additive manufacturing machine reads the sliced model and begins building the object layer by layer.
- **Post-processing**: After printing, the object may require additional steps, such as removing support structures, surface finishing, or curing.

5.3.7.2 Types of Additive Manufacturing Techniques

The brief classification of additive manufacturing is given in Figure 5.7. Additive manufacturing is classified into three main categories as liquid-based, solid-based, and powder-based additive manufacturing processes. Furthermore, these are divided into sub-processes based on the type of material, process mechanism, and the physics of the process. Fused deposition method, stereolithography, and digital light processing are based on liquid-based materials and are suitable for polymers and their composites. Lamination processes lie under solid-based processes. Powder-based processes are divided into laser sintering, laser melting, electron beam melting,

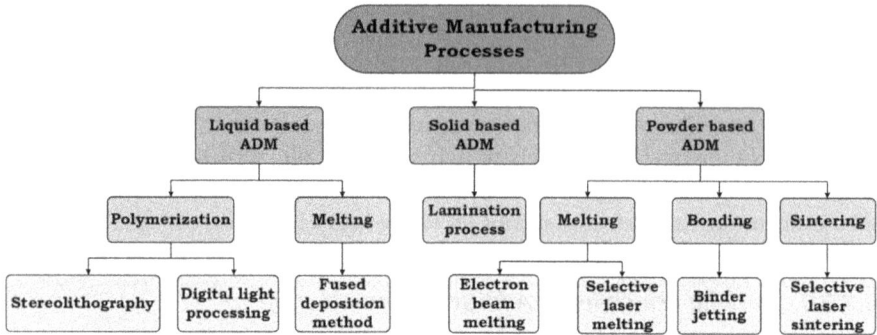

FIGURE 5.7
The classification of additive manufacturing.

and binder jetting additive manufacturing processes. Metal, ceramics, and their composites are the processing materials for these processes.

5.3.7.3 Applications of Additive Manufacturing

- **Aerospace**: It can produce lightweight, complex components for aircraft and spacecraft.
- **Healthcare**: It can create customized prosthetics, dental implants, and surgical instruments.
- **Automotive**: It is widely used in fabricating prototypes, tooling, and end-use parts.
- **Research and development**: ADM is enabled to rapid prototyping of novel research and development, which speeds up the design and testing process.
- **Consumer goods**: Able to design customized jewelry, toys, and household items.
- **Education**: It is enabling hands-on learning and rapid prototyping in classrooms.

5.3.7.4 Advantages of Additive Manufacturing

- **Freeform design**: Allows for the creation of complex geometries that are difficult or impossible with traditional methods.
- **Rapid prototyping**: Speeds up the design and testing process by quickly producing prototypes.
- **Customization**: Enables the production of personalized items tailored to individual needs.

- **Reduced waste**: Minimizes material waste by using only the necessary amount of material.
- **On-demand production**: Allows for the manufacturing of products as needed, reducing inventory costs.

5.3.7.5 Challenges of Additive Manufacturing

- **Material limitations**: A limited range of materials can be processed as compared to traditional manufacturing.
- **Surface finish**: It may require additional post-processing to achieve the desired surface quality.
- **Cost**: Initial setup and equipment costs can be high.
- **Production speed**: Generally slower than traditional manufacturing methods for large-scale production.

Additive manufacturing is revolutionizing various industries by offering new possibilities in design, customization, reverse engineering, and efficiency. Its ability to create complex, customized products with minimal waste makes it a valuable tool for modern manufacturing.

5.4 Green Manufacturing

Green manufacturing is also known as sustainable manufacturing. It refers to the creation of products using processes that minimize environmental impacts, conserve energy and natural resources, and are safe for employees, communities, and consumers. It aims to balance industrial productivity with ecological sustainability. The key concepts of green manufacturing are resource efficiency, minimizing emissions of pollutants by implementing waste treatment and reduction strategies. In addition, it is energy efficient, focuses on the utilization of renewable energy sources (solar, wind, biomass), and improves process energy efficiency with smart controls and insulation. It also comprises life cycle assessment (LCA) by evaluating the environmental impacts of a product from raw material extraction to disposal, which helps in designing eco-friendly products and processes. Green manufacturing consists of eco-design, where designing products for durability, upgradability, recyclability, and ease of disassembly is included. One important factor is focus on the utilization and development of sustainable materials, like using biodegradable, recycled, or low-impact materials (e.g., bioplastics, natural fibers) and fabricating using cleaner production techniques. These fabrication processes implement such advanced manufacturing methods that reduce hazardous substances and promote safer alternatives. This also included smart manufacturing through implementing IoT and AI to optimize energy and resource use.

5.4.1 Key Aspects of Green Manufacturing

Green manufacturing encompasses various strategies focused at reducing resource consumption, emissions, waste production, and environment eco-friendly. These are the key aspects of green manufacturing that are required to follow.

5.4.1.1 Energy Efficiency

Implementing energy-efficient technologies and practices is the central focus of green manufacturing. This includes upgrading to high-efficiency machinery, optimizing production processes, and incorporating renewable energy sources like solar or wind power to reduce overall energy consumption and emissions.

5.4.1.2 Waste Reduction and Recycling

Recycling programs and waste reduction initiatives are required to minimize waste generation through green manufacturing principles. Recycling is not only good for environment but also leads to cost savings. Strategies such as just-in-time inventory management and continuous process improvements are effective in achieving this goal.

5.4.1.3 Sustainable Material Usage

Selection of sustainable and environmentally friendly materials is a crucial task. This involves considering the lifecycle impact of materials, including their extraction, production, and disposal, and opting for materials that are recyclable, biodegradable, and eco-friendly.

5.4.1.4 Water Conservation

Water-saving measures in manufacturing processes help minimize water consumption and reduce the environmental impact on local water sources. The implementation of techniques like process water recycling, immediate leak rectification. And the use of water-efficient systems can effectively contribute to freshwater production.

5.4.1.5 Green Supply Chain Management

Sustainability throughout the supply chain can be ensured by collaborating with suppliers who follow environmental standards. This includes evaluating suppliers based on their environmental credentials and encouraging the use of recycled materials and ethical sourcing practices.

5.4.1.6 Life Cycle Assessment and Circular Economy

Evaluating the environmental impact of a product throughout its entire life cycle, from raw material extraction to disposal, to identify the areas for improvement. Life cycle assessment is indirectly associated with waste reduction and recycling, ultimately leading to cost savings. This approach aligns with the *'waste to wealth'* concept, which involves transforming discarded materials into valuable resources through sustainable and innovative methods.

Theoretical Questions and Problems

1. What are photolithography and soft lithography, and how do they differ?
2. Describe the role of deposition techniques in microfabrication processes.
3. How are MEMS devices typically fabricated?
4. Compare electrical discharge machining, electrochemical machining, and ultrasonic machining in terms of material removal rate, surface finish, and tool wear.
5. What are major considerations in selecting an advanced machining process for biomedical implants?
6. Why are non-traditional machining processes more suitable for hard, brittle, or heat-sensitive materials?
7. Discuss role of automation and CNC in integrating advanced machining processes into Industry 4.0.
8. What are the primary differences between FDM, SLA, and SLS 3D printing technologies?
9. How does layer thickness affect the mechanical properties and surface finish of 3D-printed parts?

10. Explain the role of support structures in additive manufacturing and how they are removed.

11. What are the advantages and limitations of using biopolymers in 3D printing for medical applications?

12. How does additive manufacturing compare to traditional manufacturing in terms of material waste and design freedom?

13. What is the concept of green manufacturing, and why is it important?

14. How can life cycle assessment be used to evaluate the environmental impact of a product?

15. Describe methods to reduce energy consumption in manufacturing.

16. What are biodegradable materials, and how are they integrated into fabrication processes?

17. Explain the challenges of recycling composites and advanced materials.

6

Advanced Characterization Techniques

6.1 Introduction

Characterization of materials involves analyzing their microstructure, composition, properties, and performance. Characterization techniques are widely used in various engineering domains, such as materials science, medical science, and forensics. Characterization techniques are categorized based on the application of materials. Various applications require specific properties and characteristics of the materials developed. Characterization techniques are accordingly divided into six categories as shown in Figure 6.1.

6.1.1 Structural Characterization

Crystallographic techniques like X-ray diffraction (XRD) help to determine the crystal structure of materials, while scanning electron microscopy (SEM), transmission electron microscopy (TEM), and atomic force microscopy (AFM) help reveal surface morphology and internal structure at the micro and nano-scales.

6.1.2 Chemical Characterization

Energy-dispersive X-ray spectroscopy (EDX), Fourier transform infrared spectroscopy (FTIR), and Raman spectroscopy are used to identify the chemical composition and bonding of the materials. Mass spectrometry determines the molecular composition and their structure by measuring the mass-to-charge ratio of ions.

6.1.3 Physical Characterization

Thermal analysis methods like differential scanning calorimetry (DSC) and thermogravimetric analysis (TGA) are used to measure thermal properties, such as melting point, heat capacity, and decomposition temperature. Mechanical testing methods such as tensile testing, hardness testing, and

DOI: 10.1201/9781003664871-6

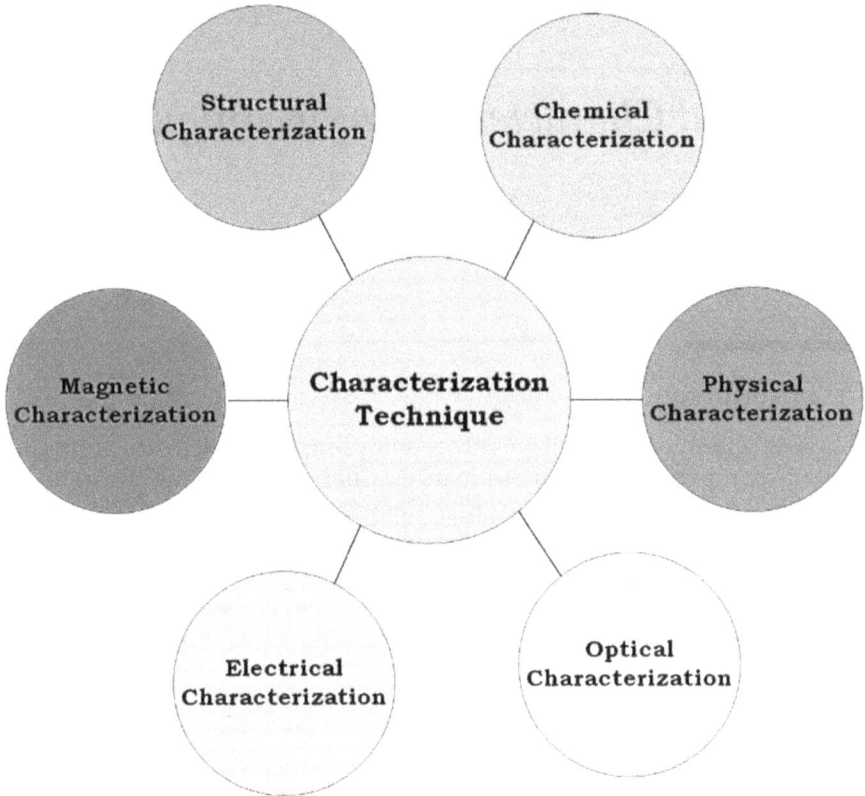

FIGURE 6.1
Classification of characterization techniques.

impact testing are used to evaluate mechanical properties like strength, hardness, and toughness, all of which are under physical characterization.

6.1.4 Optical Characterization

Light optical microscopy (LOM) and polarization microscopy provide visual analysis of material surfaces and structures. Photoluminescence spectroscopy measures the emission of light from a material after it absorbs photons, revealing the photonic, electronic, and optical properties of the material.

6.1.5 Electrical Characterization

Three/four-point probe measurements are used to determine a material's electrical conductivity. A dielectric spectroscopy analyzer is used to measure the dielectric properties of materials under varying frequencies and temperatures.

6.1.6 Magnetic Characterization

Vibrating sample magnetometry (VSM) measures magnetic properties, such as magnetization and coercivity. Magnetic resonance imaging (MRI) provides detailed images of magnetic properties.

6.2 Importance of Characterization of Materials

Characterization is the cornerstone of material science, enabling a deeper understanding of materials and their potential applications. It is a critical step in the development of innovative technologies and solutions across various fields. Characterization provides critical data on material properties, enabling informed decisions in material selection and application. Insights from characterization lead to the development of materials with enhanced performance and new functionalities. It ensures consistency and reliability in manufacturing processes by verifying material properties against standards. It also drives the discovery and development of new materials for advanced technologies and applications.

Microscopic characterization techniques are essential for analyzing the microstructure, composition, and properties of materials at the micro- and nano-scales. Here is an overview of the working principle of several advanced microscopic characterization techniques, along with instrumental details, their advantages, and applications. Each of these techniques provides unique insights into the microstructural and compositional details of materials, making them indispensable tools for scientific research and industrial applications.

6.3 Microscopy

Microscopy is a material characterization technique designed to visualize structures that are far too small for the naked eye. Its primary function is to produce a detailed image of the area under observation, whether that is within or on the surface of a sample, depending on the technique used for the characterization. The most widely used microscopy techniques are three: optical microscopy, electron microscopy, and probe microscopy. Optical microscopy uses visible light, electrons are the source of scanning the sample in electron microscopy, and movement of the probe is the excitation source in probe microscopy, respectively. These microscopy techniques are discussed in subsequent sections.

6.3.1 Light Optical Microscopy

LOM is one of the most basic and widely used techniques in material science, biology, and various other fields. The working principle and components of LOM are introduced to understand its working in depth and applications.

6.3.1.1 Working Principles

LOM involves using visible light and a system of lenses to magnify images of small objects. The lens system includes an objective lens and an eyepiece. It operates based on the principles of light, which are transmission, reflection, and refraction. LOM remains the cornerstone in scientific research and practical applications, offering valuable insights into the microscopic world. The surface finish of the material is one of the requisite factors for this microscopic analysis. It has various components, which are listed as follows.

6.3.1.2 Components

- **Light source**: Provides the illumination that is needed to observe the sample.
- **Lenses**: Include the objective lens (closest to the sample) and the eyepiece lens (through which the user views). These lenses work together on principles of light to magnify the image.
- **Sample holder**: The platform where the sample is placed for observation and is positioned between the light source and the objective lens.
- **Focus mechanism**: Adjusts the distance between the sample and the objective lens to bring the image into focus.
- **Condenser**: Focuses light onto the sample, thereby improving illumination and contrast.

6.3.1.3 Applications

- **Biological sciences**: It is used to observe cells, tissues, and microorganisms.
- **Material science**: This enables the study of material structures, including metals, polymers, and composites.
- **Medical diagnostics**: This type of microscopy assists in identifying pathogens and analyzing tissue samples.
- **Forensics**: Helps in examining evidence such as fibers and residues.

6.3.1.4 Advantages

- **Non-destructive**: Samples can often be examined without destroying them.
- **Versatile**: It can be used with a variety of samples, including biological tissues and industrial materials.
- **Accessible**: Relatively simple and affordable as compared to more advanced techniques.

6.3.1.5 Types of Light Optical Microscopy

- **Brightfield microscopy**: This is the simplest form of LOM, where light passes through the sample, and the image is formed by absorption or reflection.
- **Darkfield microscopy**: This type of LOM enhances contrast in unstained samples; light is directed at an angle, and only scattered light is observed.
- **Phase contrast microscopy**: This converts phase shifts in the light passing through transparent specimens into variations in intensity, thereby enhancing contrast. This is a vital tool for the clinical field and provides enhanced contrast and detail of specimens that are generally challenging to observe.
- **Fluorescence microscopy**: Uses fluorescent dyes that emit light upon excitation, allowing the observation of specific features within a sample. This microscopy is widely used in cell culture studies and biological applications.

6.3.2 Polarization Microscopy

Polarization microscopy is a powerful optical microscopy technique that enhances the contrast of specimens that are birefringent (i.e., materials that split light into two rays when it passes through them). It is widely used in various scientific fields to study the optical properties, composition, and structural details of materials.

6.3.2.1 Working Principles

Polarization microscopy uses polarized light (vibrations of the electromagnetic waves are restricted to a particular direction) to observe and analyze birefringent materials. When polarized light passes through such materials, it experiences changes in its velocity and direction, resulting in unique optical effects that can be captured and analyzed. This microscope uses two filters for generating images; one is a polarizer and the other is an analyzer. Light emitted from the source is polarized after passing through a polarizer and falls over the sample, passing through a condenser lens. Furthermore, the sample's internal structures influence the polarized light and form an image on the camera after passing through an objective lens and analyzer.

6.3.2.2 Components

- **Polarizer**: It is positioned below the sample. It polarizes the incoming light.
- **Analyzer**: It is positioned above the sample. It analyzes the light that has passed through the sample.
- **Rotatable stage**: Allows the sample to rotate to view it from different angles.
- **Condenser and objective lenses**: Focus light coming from the light source onto the sample and magnify the image, respectively.

6.3.2.3 Types of Polarization Microscopy

- **Crossed polarization**: In this type, both polarizer and analyzer are oriented perpendicular to each other, enhancing contrast for birefringent materials.
- **Conoscopic polarization**: This uses a condenser with a high numerical aperture to observe interference patterns; it is useful in crystallography.

6.3.2.4 Applications

- **Geology**: Used in identifying and studying minerals and rocks based on their birefringent properties.
- **Biology**: Used in observing the structure of biological tissues and fibers, such as muscle and collagen.
- **Material science**: Used to analyze polymers, liquid crystals, and other synthetic materials.
- **Forensics**: Used to examine fibers, pigments, and other materials with distinct birefringent characteristics.

6.3.2.5 Advantages

- **Enhanced contrast**: Provides high-contrast images of birefringent materials, revealing details that are not visible using standard optical microscopy.

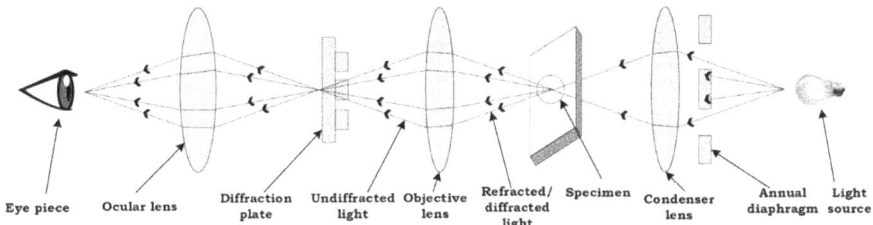

FIGURE 6.2
Schematic of phase contrast microscopy and its working principles.

- **Non-destructive**: Allows for observation of materials without damaging them.
- **Versatile**: Applicable to a wide range of materials and scientific disciplines.

6.3.3 Phase Contrast Microscopy

Phase contrast microscopy is a technique that enhances the contrast of transparent and colorless specimens, making it easier to observe details that would otherwise be difficult to see with standard brightfield microscopy. This technique is an invaluable tool for researchers and clinicians, providing enhanced contrast and details of specimens that are otherwise challenging to observe. It is particularly useful in fields where observing live cells and natural specimens is essential.

6.3.3.1 Working Principles

Phase contrast microscopy converts phase shifts in the light passing through a transparent specimen into differences in intensity. These phase shifts occur because different parts of the specimen have different refractive indices, causing light to slow down and change phase as it passes through. A small phase change in light creates a change in brightness, which can be observed in the image of the sample. A diffractive lens is positioned after the objective lens, creating an interferogram, containing phase information that is useful to reconstruct the surface topography of the surface. Figure 6.2 shows the various components of phase contrast microscopy.

6.3.3.2 Components

- **Light source**: Provides illumination for the specimen.
- **Condenser**: It includes a phase ring that partially shifts the phase of the light.

- **Objective lenses**: Contain a phase plate that further enhances the phase shift, converting it into intensity differences.
- **Specimen stage**: It holds the specimen in place for observation.
- **Diffractive lens**: It creates an interferogram that possesses phase information, which is further used to reconstruct the surface morphology of the surface.
- **Eyepiece**: Allows the viewer to see the enhanced contrast image of the sample.

6.3.3.3 Applications

- **Cell biology**: It is used in observing live cells and tissues without the need for staining, preserving their natural state.
- **Microbiology**: Studying microorganisms, such as bacteria and protozoa, which are typically transparent and difficult to see under brightfield illumination.
- **Medical diagnostics**: Examining unstained biological specimens, including blood cells and tissue samples, for diagnostic purposes.
- **Material science**: Investigating thin films, fibers, and other transparent materials.

6.3.3.4 Advantages

- **Non-destructive**: Allows for the observation of live cells and unstained specimens, preserving their natural state.
- **Enhanced contrast**: Provides high-contrast images of transparent and colorless specimens that are difficult to see with standard brightfield microscopy.
- **Ease of use**: Relatively simple to set up and use with standard light microscopes equipped with phase contrast components.

6.3.3.5 Limitations

- **Light requirements**: More light is needed for phase contrast than for brightfield microscopy.
- **Artifact introduction**: The phase contrast technique can sometimes introduce halo artifacts around the edges of structures that damage the image and create errors.
- **Limited to transparent specimens**: Most effective with transparent and colorless specimens; less useful for opaque or stained samples.
- **Distorted images**: Thick specimens can appear distorted.

6.3.4 Differential Interference Contrast Microscopy

Differential interference contrast (DIC) microscopy is also known as Nomarski microscopy. This is an advanced optical microscopy technique that enhances contrast in unstained, transparent specimens by using differences in refractive index to produce detailed, three-dimensional-like images. Its ability to reveal fine details and edges makes it particularly useful for studying live cells and delicate structures. This microscopy is used in cell biology, neuroscience, and material science to study nano features.

6.3.4.1 Working Principles

DIC microscopy works by splitting a beam of polarized light into two beams that pass through the specimen at slightly different angles. These beams are then recombined before reaching the observer. Any differences in the refractive index within the specimen cause the two beams to interfere with each other, creating a high-contrast image that highlights edges and fine details.

6.3.4.2 Components

- **Polarizer**: A polarizer is a type of optical filter that polarizes the incoming light before it reaches the specimen.
- **Nomarski (Wollaston) prism**: A prism is used to split the polarized light into two beams with slightly different paths to create a 3-D image of the specimen.
- **Objective lenses**: The persistence of an objective lens is to focus the split beams through the specimen and onto the detector.
- **Analyzer**: Recombines the two beams after they pass through the specimen, creating the interference contrast.
- **Differential interference filter**: Enhances the contrast by adjusting the phase relationship between the two beams.

6.3.4.3 Applications

- **Cell biology**: It is widely used in observing live, unstained cells and cellular structures with enhanced contrast and detail.
- **Developmental biology**: To study embryos and other transparent specimens for advancement in biological research.
- **Neuroscience**: To examine fine details in neural tissues and cells.
- **Material science**: Analyzing the surface features of transparent materials and thin films.

6.3.4.4 Advantages

- **Enhanced contrast**: Provides high-contrast, three-dimensional-like images of transparent specimens without the need for staining.
- **Non-destructive**: Allows for the observation of live cells and natural specimens, preserving their original state.
- **Detailed imaging**: Offers fine detail and edge enhancement, making it easier to observe structures that are difficult to see with standard microscopy.

6.3.4.5 Limitations

- **Complex setup**: Requires specialized equipment and precise alignment of optical components.
- **Artifact introduction**: Introduces optical artifacts if not properly adjusted.
- **Limited to transparent specimens**: Most effective with transparent and unstained samples; less useful for opaque or heavily stained specimens.

6.3.5 Confocal Microscopy

Confocal microscopy is an advanced optical imaging technique that was developed in the 1950s at Harvard University for neuroscience research by Marvin Minsky. This technique provides high-resolution and high-contrast images of specimens by eliminating out-of-focus light. It can be used for thick samples without any distortion. It is widely used in biology, medical science, and material sciences for detailed imaging of cells, tissues, and materials. An appropriate discussion on confocal microscopy is presented with the working principles.

6.3.5.1 Working Principles

Point illumination is the key principle in confocal microscopy. This working mechanism uses a spatial pinhole to create point illumination that eliminates out-of-focus light in specimens that are thicker than the focal plane. This results in clear and sharp images of specific optical sections within a sample. This is influenced by the objective lens, which is a measure to gather light and resolve fine specimen detail at a fixed object distance. The depth of the illumination point is about 0.6–1.5 µm in confocal microscopy. This depth affects how much of the sample is in focus at one time and is also related to the depth of the field of the microscope.

The brightest intensity occurs at the focal point, where the light is most concentrated.

6.3.5.2 Components

- **Laser light source**: Provides high-intensity, coherent light for illuminating the sample.
- **Scanning system**: This consists of dichroic mirrors or a galvo system to scan the laser beam across the specimen.
- **Objective lens**: Focuses the laser beam onto the specimen and collects emitted light from the sample.
- **Confocal pinhole**: Removes out-of-focus light, allowing only light from the focal plane to reach the detector.
- **Detector**: Captures in-focus light to create an image and is used in detectors including photomultiplier tubes (PMTs) and charge-coupled devices (CCDs).
- **Computer**: Processes and reconstructs the scanned data into a digital image.

6.3.5.3 Applications

- **Cell biology**: It is used in observing the detailed structure and dynamics of live cells and tissues.
- **Neuroscience**: Imaging neural networks and brain slices with high resolution.
- **Developmental biology**: Used in studying the development of embryos and organogenesis for advanced research.
- **Material science**: Analyzing the surface and internal structure of materials, including polymers and composites.

6.3.5.4 Advantages

- **Live imaging**: Enables real-time observation of dynamic processes in living cells and tissues.
- **Optical sectioning**: Allows for the imaging of specific planes within a specimen, providing depth information and 3-D reconstruction.
- **High resolution**: Delivers sharp and detailed images by reducing background noise and out-of-focus light.
- **Multicolor imaging**: It can be used with multiple fluorescent dyes to study different components of a specimen simultaneously.

6.3.5.5 *Limitations*

- **Photobleaching**: Prolonged exposure to intense laser light can cause photobleaching of fluorescent dyes, reducing signal intensity over time.
- **Limited penetration depth**: Effective for imaging specimens with limited thickness; not suitable for very thick or opaque samples.
- **Complex setup**: Requires precise alignment and calibration of optical components, as well as specialized software for image processing.

Confocal microscopy is a versatile imaging technique that provides detailed insights into the structure and function of biological and material specimens.

6.3.6 Scanning Electron Microscopy

SEM is a highly versatile imaging technique that is used to analyze the surface topography, morphology, and composition of specimens at high resolution. SEM is a type of electron microscopy that is used to create an image of a sample by scanning the sample's surface with a high-voltage focused beam of electrons. It provides detailed images and is widely used in various fields, including materials science, biology, and engineering.

6.3.6.1 *Working Principles*

SEM works by scanning a focused beam of electrons across the surface of the sample. When the electrons interact with the atoms in the sample, they produce different kinds of signals that are detected and used to create an image. These signals include secondary electrons, backscattered electrons, characteristic X-rays, Auger electrons, and other rays from different shells of atoms. Each entity provides different types of information about the sample. Figure 6.3(a) depicts the pear model of the electron beam interaction with the sample.

  In SEM, the pear shape defines the interaction volume, where the electron beam interacts with the sample. The shape of the pear depends on the density of the sample material and the electron beam voltage. Secondary electrons are emitted from the nearest surface of the sample when the electron beam interacts with the sample and provide morphological information about the sample. When an electron beam hits the sample's surface, some electrons are scattered from the surface and these are backscattered electrons. These electrons are high-energy electrons that belong to the electron beam. The last important attribute is the characteristic X-rays that are emitted from the depth of the sample, and they provide compositional information. These X-rays are used for EDX analysis. Furthermore, the schematic diagram of SEM and microimage of silver nanoparticles are shown in Figure 6.3(b, c).

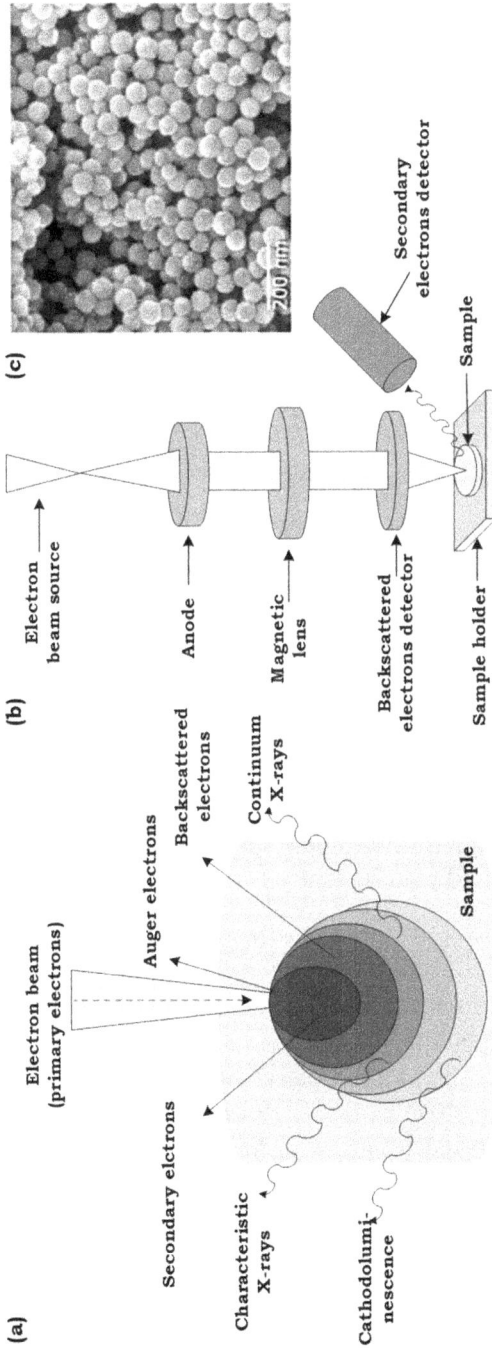

FIGURE 6.3
(a) Pear model of electron beam interaction with the sample in SEM, (b) schematic of SEM, and (c) microimage of nanoparticles.

6.3.6.2 Components

- **Electron gun**: Generates a focused beam of electrons using a tungsten gun.
- **Electromagnetic lenses**: They are used to focus and direct the electron beam onto the specimen.
- **Scanning coils**: Move the electron beam across the sample in a raster pattern.
- **Detectors**: Capture the emitted signals, such as secondary electrons and backscattered electrons, to form a microimage.
- **Specimen chamber**: Holds the sample in a vacuum environment to prevent electron scattering by air molecules.
- **Control system**: Allows the user to adjust parameters and view the resulting images on a computer screen.

6.3.6.3 Applications

- **Materials science**: It analyze the surface structure, composition, and properties of metals, ceramics, polymers, and composites.
- **Biology**: To observe the detailed surface morphology of cells, tissues, and microorganisms.
- **Forensics**: To examine evidence such as gunshot residues, fibers, and other trace materials.
- **Semiconductor industry**: To inspect the surface features and defects of microelectronic devices.
- **Nanotechnology**: To investigate the morphology and structure of nanomaterials and nanostructures.

6.3.6.4 Advantages

- **High resolution**: Provides detailed images with high resolution down to a few nanometers.
- **3-dimensional imaging**: Produces images that appear three-dimensional, offering valuable insights into surface topography.
- **Elemental analysis**: It can be combined with EDX for elemental composition analysis.
- **Versatility**: Applicable to a wide range of materials and samples, both conductive and non-conductive (with appropriate preparation).

6.3.6.5 Limitations

- **Sample preparation**: Non-conductive samples often require coating with a thin layer of conductive material, such as gold or carbon, to prevent charging.
- **Vacuum requirement**: Samples must be placed in a vacuum chamber, which can limit the types of specimens that can be analyzed.
- **Damage risk**: High-energy electron beams can potentially damage delicate samples.

6.3.7 Transmission Electron Microscopy

TEM is one of the excellent electron beam techniques that is used to investigate the microscopic nature of defects such as dislocations and impurities. This technique allows scientists to observe the internal structure of materials at the atomic level. It also uses a high-energy electron beam to form an image of a very small object. It has a higher resolution than any other microscopic imaging technique and provides high-resolution images and detailed information about the composition, crystallography, and morphology of the sample.

6.3.7.1 Working Principles

TEM works by transmitting a high-energy electron beam through an ultrathin sample. As the electrons pass through the sample, they interact with the atoms and produce various signals that can be detected to form an image. As electrons have much shorter wavelengths than visible light, TEM can achieve much higher resolution than optical microscopy. The sample is most often an ultrathin section, less than ~100 nm thick, or a suspension on a grid. In Figure 6.4(a), the transmission of electrons through the sample is depicted.

There is elastic scattering, which has electrons of the same energy that they possessed before the interaction with the sample in different directions. Elastically scattered electrons are the main source to create contrast in the TEM microimage. Inelastic scattered electrons lose energy and direction. Due to less energy, these electrons influence the image of the sample. Furthermore, incoherent elastic scattered electrons help in visualizing the sample and create a magnified image. Figure 6.4(b–d) displays the schematic diagram of TEM and images of silver nanoparticles, and an image of bacteria extracted from TEM, respectively.

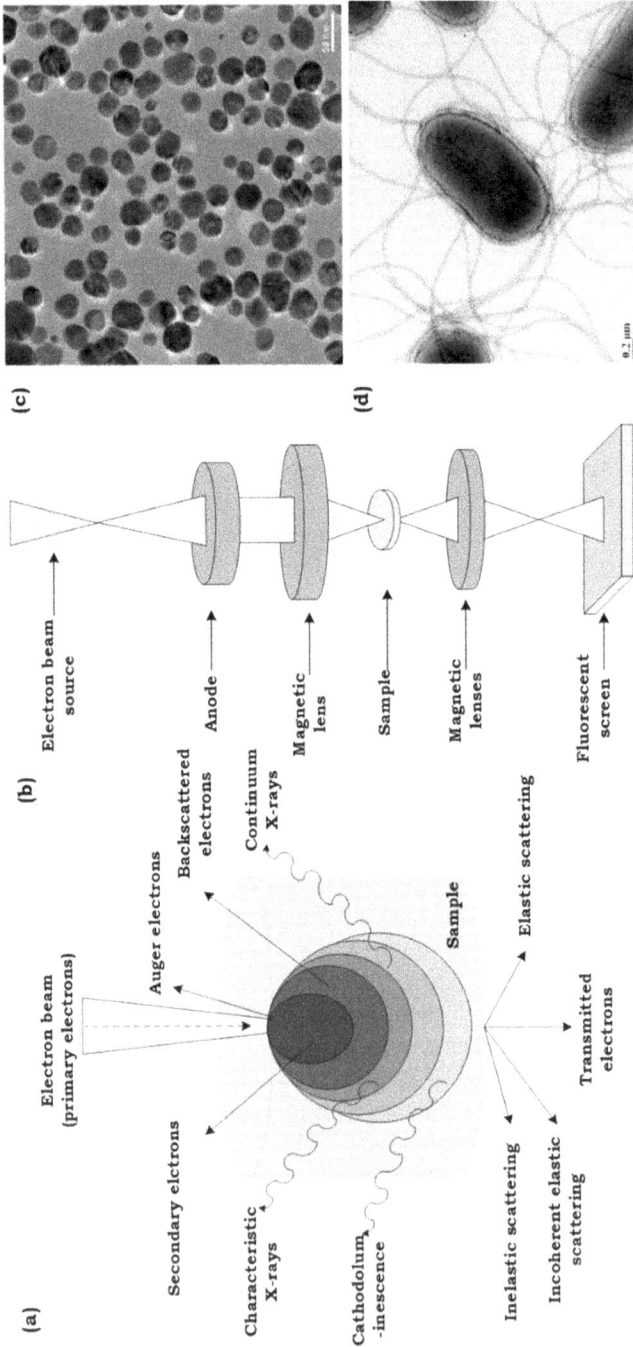

FIGURE 6.4

(a) A pear model of electron beam interaction with the sample in TEM, (b) schematic diagram of TEM, (c) TEM image of silver nanoparticles, and (d) TEM image of bacteria.

6.3.7.2 Components

- **Electron gun**: Generates a focused beam of electrons of high energy.
- **Condenser lenses**: Focus and control the electron beam before it interacts with the sample.
- **Specimen holder**: Holds the ultra-thin specimen in place.
- **Objective lens**: Magnifies the image and focuses the electrons that have passed through the specimen.
- **Intermediate and projector lenses**: Further magnify and project the image onto a detector or viewing screen.
- **Detector/Camera**: Captures the transmitted electron image for analysis and recording.

6.3.7.3 Applications

- **Materials science**: This is important for analyzing the internal structure of metals, alloys, ceramics, and nanomaterials.
- **Biology**: To observe the ultrastructure of cells, viruses, and tissues.
- **Semiconductor industry**: It is widely used for inspecting the internal features of microelectronic devices.
- **Nanotechnology**: This is used to investigate the structure and properties of nanoparticles and nanocomposites.

6.3.7.4 Advantages

- **High resolution**: TEM can achieve resolutions down to the atomic level, allowing for the visualization of individual atoms.
- **Detailed structural information**: It can provide insights into the crystallographic structure, defects, and composition of materials.
- **Versatility**: It can be used to study a wide range of materials, including biological specimens, metals, and semiconductors.

6.3.7.5 Limitations

- **Sample preparation**: It requires the preparation of ultra-thin specimens, typically less than 100 nanometers thick, which is a challenging and time-consuming job.
- **Vacuum requirement**: The sample must be placed in a high vacuum environment to prevent electron scattering; consequently, a high vacuum system is required.

- **Potential damage**: High-energy electron beams can potentially damage delicate biological specimens.
- **Skilled operator and high cost**: It requires a skilled operator and a
  relatively higher installation cost compared to other characterization
  techniques.

6.3.8 Atomic Force Microscopy

AFM is a high-resolution imaging technique that allows scientists to visualize and measure the surface topography of materials at the nanoscale. The
working principle of AFM is to measure the force between a sharp scanning
probe and the sample's surface at the atomic level. It provides 3-D images
using the value of atomic force and is widely used in materials science, biology, and nanotechnology.

6.3.8.1 Working Principles

AFM works by scanning a sharp probe set on a cantilever beam across the surface of a sample. The probe interacts with the surface forces, causing the cantilever beam to deflect. These deflections are measured and used to construct
an image of the surface with atomic or near-atomic resolution. The laser beam
reflects off due to the bending of the cantilever beam and is detected by a photodetector. The photodetector converts the light coming from the laser beam
into an electrical signal and is further used to form a 3-D image of the sample's
surface on the computer screen. Figure 6.5(a) shows the schematic diagram of
AFM with its working principle. The micrographs of the Coronavirus are displayed in Figure 6.5(b) extracted from the AFM. These images are 3-D images
of different colors that are selected by the software of the AFM image analysis.

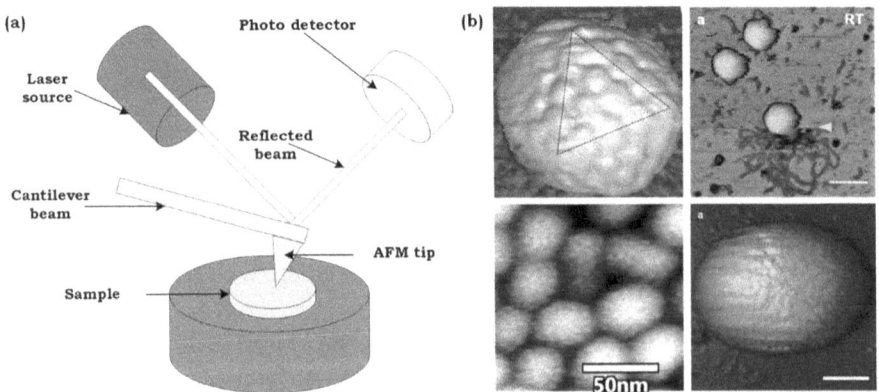

FIGURE 6.5
(a) Schematic diagram of AFM and (b) microimages of the SARS-CoV-19 virus extracted from AFM.

6.3.8.2 Components

- **Cantilever and probe**: A flexible cantilever has a sharp probe at its tip. The probe scans the surface of the specimen.
- **Piezoelectric scanner**: It moves the cantilever and probe precisely over the sample in three dimensions.
- **Laser and photodetector**: A laser beam reflects off the back of the cantilever and onto a photodetector, measuring the deflections of the cantilever.
- **Controller**: It processes the deflection signals and generates a topographic image of the specimen surface.

6.3.8.3 Modes of Operation

- **Contact mode**: The probe remains in constant contact with the surface, providing high-resolution images but potentially causing damage to samples of soft materials.
- **Tapping mode**: A probe oscillates near the surface, tapping it gently. This mode reduces damage to the sample and provides high-resolution images.
- **Non-contact mode**: A probe oscillates above the surface without making contact, measuring attractive forces between the probe and the sample. This mode is suitable for very soft or delicate samples.

6.3.8.4 Applications

- **Materials science**: It is used in analyzing the surface structure, roughness, and mechanical properties of metals, polymers, ceramics, and composites.
- **Biology**: It is used to observe the topography of cells, proteins, DNA, and other biological molecules.
- **Nanotechnology**: It is used to investigate nanostructures, nanoparticles, and nanocomposites.
- **Semiconductors**: It is used in examining the surface features and defects of microelectronic devices.

6.3.8.5 Advantages

- **High resolution**: Provides images with atomic or near-atomic resolution, revealing fine surface details.
- **3-dimensional imaging**: Produces three-dimensional topographic images of the specimen surface.

- **Versatility**: Applicable to a wide range of materials, including conductive and non-conductive specimens.
- **Minimal sample preparation**: Requires little to no sample preparation, preserving the natural state of the specimen.

6.3.8.6 Limitations

- **Limited scan size**: The scan area is relatively small, typically on the order of micrometers.
- **Slow imaging speed**: Scanning can be time-consuming, especially for large areas or high-resolution images.
- **Sample damage**: Contact mode can damage soft or delicate samples due to the interaction between the probe and the surface.

6.4 Standard Metallographic Sample Preparation Techniques

Sample preparation in metallographic study is a critical process that involves several steps to ensure that the microstructure of a material is accurately revealed for examination under a microscope. Proper preparation is essential to avoid altering the true microstructure and to obtain reliable results. Figure 6.6 demonstrates the various steps that are required to follow in sample preparation for microstructural study. A detailed understanding of

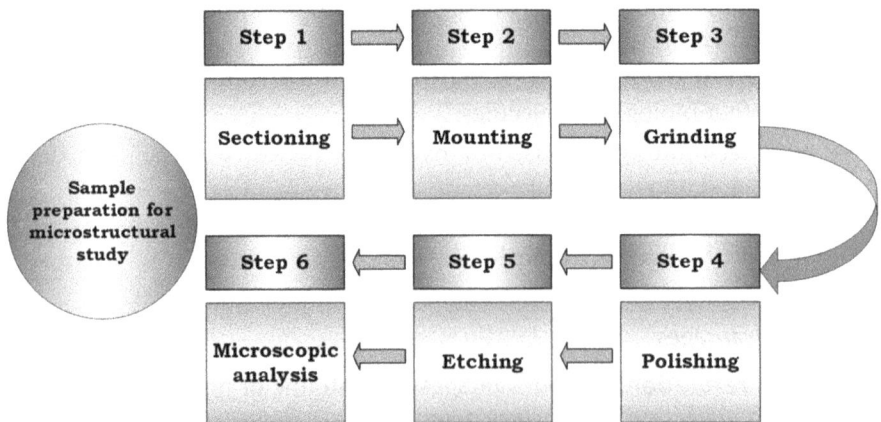

FIGURE 6.6
Procedure for sample preparation for microstructural study.

each step is required before starting to follow the steps mentioned. There are some international standards for sample preparation of various materials for microstructural analysis. These standards are designed by the ASTM and there are some standards such as *E3*, *E407*, *E768*, *E883*, and *E1951* for various steps of sample preparation of different materials to obtain microstructures.

1. **Sectioning**: It is the first step that involves cutting a required-size sample from the material. This is typically done using abrasive/diamond cutting techniques to minimize damage and ensure a smooth surface. Generally, a diamond cutting wheel is used for sectioning the metastable materials for metallographic study.

2. **Mounting**: The cut sample is then mounted in a mounting press using a mounting material (e.g., epoxy resin) to provide stability and ease of handling during subsequent steps. Two types of mounting are used in sample preparation, such as cold mounting and hot mounting. Cold mounting is generally preferred for heat and pressure-sensitive samples, such as coated samples and PCBs.

3. **Grinding**: The mounted sample is ground using coarse abrasive papers (SiC grinding paper) to remove any surface damage and to achieve a flat surface. This step is performed in multiple stages, starting with coarse grit and progressing to finer grits. Finer grit paper provides a smoother surface.

4. **Polishing**: After grinding, the sample is polished to achieve a mirror-like finish. Polishing is done using finer abrasives and a polishing cloth or wheel to remove any remaining scratches and to prepare the surface for microscopic examination.

5. **Etching**: The polished sample is then etched using a chemical solution to reveal the microstructure. Etching selectively corrodes the surface (especially the grain boundaries), highlighting different phases, grain boundaries, and other features.

6. **Microscopic examination**: Finally, the etched sample is examined under a microscope to analyze its microstructure. Advanced techniques such as SEM and TEM may be used for more detailed analysis in terms of surface morphology, grain structure, chemical compositions, crystal structure, grain orientation, etc.

6.5 X-Ray Diffraction Techniques

XRD is a material characterization technique that is used to determine the crystallographic structure, chemical composition, atomic structure, and

physical properties of materials. This is a non-destructive characterization technique and is widely employed in various scientific fields, including materials science, chemistry, geology, physics, and forensic science. This
technique is based on the principle of diffraction of X-rays, where X-rays
penetrate the sample and diffract from the sample at the same angle (inclination angle) and extract the valuable information.

6.5.1 Working Principles

XRD is a versatile non-destructive characterization technique, based
on the diffraction of X-rays at the sample. These rays interact with the
atomic planes in the crystal lattice of the material, and they are diffracted
at specific angles (same as the angle of inclination). By measuring these
diffraction angles and intensities, one can determine the crystal structure and other properties of the material. The working mechanism of the
XRD instrument follows Bragg's law. This law defines that when incident
X-rays hit a crystal at a specific angle and wavelength, intense X-rays
reflect from the crystal carrying with them the information about the
material. The working mechanism of XRD is represented in Figure 6.7.
Bragg's law is used to analyze the crystal lattice, defects, and various
phases. The following formula is used to calculate the lattice parameters
of the crystal lattice.

$$n\lambda = 2d \times \sin\theta$$

where λ is the wavelength, d is the distance between two adjacent crystal
planes, θ is the diffraction angle, and n is a constant ($n = 1, 2, 3,...$).

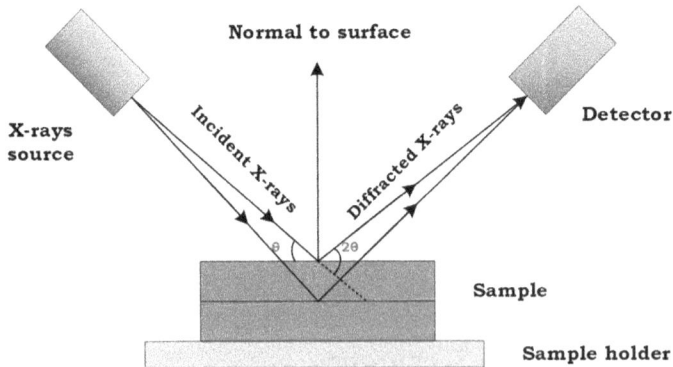

FIGURE 6.7
Schematic representation of the working principles of XRD.

6.5.2 Components

- **X-ray source**: A tungsten target with a high-voltage supply is used as a source of X-rays that is in a sealed tube or a rotating anode, generating X-rays.
- **Sample holder**: Holds the material sample to be analyzed.
- **Detector**: Captures the diffracted X-rays and measures their intensity and angle.
- **Goniometer**: Precisely positions and rotates the sample and detector to measure diffraction patterns as per the requirement.

6.5.3 Applications

- **Crystallography**: It is used for determining the atomic and molecular structure of crystals.
- **Phase identification**: It is used to identify and quantify the phases present in a material sample.
- **Stress/strain analysis**: To measure the stress and strain in crystalline materials.
- **Texture analysis**: To understand the preferred orientation of grains in polycrystalline materials.
- **Thin-film analysis**: To evaluate the composition and quality of thin films.

6.5.4 Advantages

- **Non-destructive**: Allows for the analysis of materials without causing damage to the material.
- **Precise and accurate**: Provides detailed information about the crystal structure and composition.
- **Versatile**: Applicable to a wide range of materials, including metals, ceramics, polymers, and biological samples.

6.5.5 Limitations

- **Sample preparation**: Requires samples to be finely ground and homogenous to obtain accurate results.
- **Limited to crystalline materials**: Most effective for materials with a well-defined crystal structure; less useful for amorphous or non-crystalline substances.
- **Instrumentation cost**: It requires specialized and often expensive equipment.

XRD is a fundamental technique that is used in many scientific disciplines, offering critical insights into the structure and properties of materials. Its ability to precisely determine crystallographic details makes it an invaluable tool for researchers and scientists.

6.6 Spectroscopic Characterization

Spectroscopic characterization involves the study of the interaction of electromagnetic radiation (light) with matter across various wavelengths. This technique helps identify the composition, structure, and physical properties of materials. Spectroscopy works by measuring the intensity of light as a function of wavelength or frequency following its interaction with a sample. Different materials absorb, emit, or scatter light in unique ways, providing a "fingerprint" that can be used for the identification and analysis of the materials/chemical compounds.

Steps in Spectroscopic Analysis

- **Sample preparation**: A sample is prepared and placed in the sample holder.
- **Light interaction**: A light source illuminates the sample, and the light interacts with it.
- **Signal detection**: A detector captures the light that is transmitted, absorbed, emitted, or scattered by the sample.
- **Data analysis**: A computer processes the detected signals to generate spectra, which are then interpreted to provide information about the sample's properties.

Main Types of Spectroscopy
1. Absorption Spectroscopy
2. Emission Spectroscopy
3. Fluorescence Spectroscopy
4. Raman Spectroscopy

Spectroscopic characterization is a versatile analytical tool used across various scientific disciplines to gain detailed insights into the composition and behavior of materials. Its ability to provide precise and accurate information makes it indispensable for research and industrial applications.

6.6.1 Absorption Spectroscopy

It is used to identify and quantify substances based on the way they absorb electromagnetic radiation. It measures the absorption of light by a substance at different wavelengths and frequencies. When light passes through a sample, certain wavelengths are absorbed by the sample's molecules, depending on their structure and composition. The absorbed wavelengths correspond to specific energy levels in the molecules, providing valuable information about the sample. Absorption spectroscopy is of two types, one is ultraviolet-visible-near infrared spectroscopy (UV-Vis-NIR), and the second is the FTIR. These types of spectroscopic techniques are used for chemical analysis to identify and quantify substances. These are widely used in biology to study biomolecules like proteins and DNA, and in environmental science for detecting pollutants in air and water. It is also very useful in diagnosing diseases and monitoring patient samples in medical science. The working principles of UV-Vis-NIR and FTIR spectroscopies are discussed in detail in the following section.

6.6.1.1 Ultraviolet-Visible-Near Infrared Spectroscopy

UV-Vis-NIR is a comprehensive analytical technique that encompasses the ultraviolet (UV), visible (Vis), and near-infrared (NIR) regions of the electromagnetic spectrum. This versatility allows scientists to analyze a wide range of samples with a single instrument. It works in wide range of regions such as ultraviolet, visible and near-visible.

- **Ultraviolet region**: Typically ranges from 10 to 400 nanometers (nm). It is useful for detecting and studying electronic transitions in molecules and atoms.
- **Visible region**: Ranges from 400 to 700 nm. This range is associated with colors visible to the human eye and is often used for studying the absorption and transmission of light in various substances.
- **Near-infrared region**: Ranges from 700 to 2,500 nm. It is particularly useful for studying molecular vibrations and overtones, providing valuable information about the chemical composition and structure of samples.

6.6.1.1.1 Components
- **Light source**: Emits a broad spectrum of light; UV, visible, and NIR regions.
- **Monochromator**: Separates the light into its component wavelengths.
- **Sample holder**: Contains the sample for analysis.

- **Detector**: Measures the intensity of light at different wavelengths.
- **Data analysis system**: Processes and records the spectral data.

6.6.1.1.2 Working Principles

- **Light emission**: A broad-spectrum light source emits light across the UV, visible, and NIR regions.
- **Wavelength selection**: The monochromator selects specific wavelengths of light.
- **Sample interaction**: Selected light passes through or reflects off the sample. Specific wavelengths are absorbed depending on the sample's molecular composition.
- **Detection**: The detector measures the intensity of the transmitted or reflected light at each wavelength.
- **Data analysis**: The resulting spectrum is analyzed to determine the sample's properties.

6.6.1.1.3 Applications

UV-Vis-NIR spectroscopy is widely used in the following:

- **Chemical analysis**: Identifying and quantifying substances in a sample.
- **Pharmaceuticals**: It is used in quality control, formulation analysis, and drug development.
- **Environmental monitoring**: Detecting pollutants and contaminants in air, water, and soil.
- **Food and agriculture**: Analyzing the quality and composition of food products.
- **Material science**: Studying the properties and structure of various materials, including polymers, coatings, and composites.

6.6.1.1.4 Advantages

- **Versatility**: Combines UV, visible, and NIR regions into one technique.
- **Non-destructive**: Non-invasive and requires minimal sample preparation.
- **Speed and precision**: Rapid data acquisition with high precision and reproducibility.

6.6.1.2 Fourier Transform Infrared Spectroscopy

FTIR involves measuring the absorption of infrared light by a sample. The unique aspect of FTIR is that it uses an interferometer to collect all wavelengths of the infrared spectrum simultaneously, and then applies a mathematical transformation (Fourier transform) to obtain the absorption spectrum. This allows for rapid and precise analysis of the sample.

6.6.1.2.1 Components

- **Infrared light source**: Emits a broad spectrum of infrared light.
- **Interferometer**: Splits the infrared light into two beams, which then recombine to produce an interference pattern.
- **Sample holder**: Contains the sample being analyzed.
- **Detector**: Measures the intensity of the transmitted or reflected infrared light after it passes through the sample.
- **Computer/Processor**: Performs the Fourier transform to convert the interference pattern into an absorption spectrum.

6.6.1.2.2 Working Principles

- **Interferometer operation**: Infrared light passes through an interferometer, creating an interference pattern.
- **Sample interaction**: Interference pattern interacts with the sample, and specific wavelengths are absorbed based on the sample's molecular structure.
- **Detection and transformation**: A detector measures the intensity of the transmitted or reflected light, and the Fourier transform converts the interference pattern into an absorption spectrum.

6.6.1.2.3 Applications

FTIR spectroscopy is used in the following:

- **Chemistry**: Identifying and quantifying chemical compounds.
- **Pharmaceuticals**: Quality control and formulation analysis.
- **Environmental science**: Detecting and analyzing pollutants.
- **Materials science**: Studying polymers, coatings, and composites.
- **Forensics**: Investigating substances found at crime scenes.

6.6.1.2.4 Advantages

- **Speed**: Rapid data acquisition due to simultaneous measurement of all wavelengths.
- **Accuracy**: High precision and reproducibility of results.
- **Versatility**: Can analyze solids, liquids, and gases with minimal sample preparation.

6.6.2 Emission Spectroscopy

Emission spectroscopy involves measuring the light emitted by a substance when it is excited by an external energy source. This excitation causes the electrons in the atoms or molecules to jump to higher energy levels. When these electrons return to their original energy levels, they release energy in the form of light. The emitted light can be analyzed to determine the composition and properties of the substance.

6.6.2.1 Components

- **Energy source**: Provides the energy to excite the atoms or molecules. This can be in the form of heat (flame), electrical discharge, or a laser.
- **Sample**: The substance being analyzed.
- **Monochromator**: Selects specific wavelengths of the emitted light for measurement.
- **Detector**: Measures the intensity of the emitted light at different wavelengths.
- **Computer/Recorder**: Analyzes and records the emission data.

6.6.2.2 Applications

Emission spectroscopy is widely used in various applications in both research and industry. Inductively coupled plasma spectroscopy is a type of emission spectroscopic technique (ICP-OES) that uses plasma to excite the sample, providing high sensitivity and precision. This technique is essential for understanding the elemental composition of substances.

- **Chemical analysis**: Identifying and quantifying elements and compounds.
- **Astronomy**: Analyzing the composition of stars and other celestial bodies.
- **Environmental science**: Detecting trace elements and pollutants in various samples.
- **Forensics**: Investigating the composition of materials found at crime scenes.

6.6.3 Fluorescence Spectroscopy

Fluorescence spectroscopy involves measuring the fluorescence emitted by a substance when it absorbs light (typically ultraviolet or visible light) and then re-emits light at a longer wavelength. This technique is based on the principle that certain molecules, called fluorophores, can absorb light at one wavelength (excitation) and emit light at a different wavelength (emission).

6.6.3.1 Components

- **Light source**: A laser or xenon lamp emits light at specific wavelengths.
- **Sample**: Substance containing fluorophores and it is measured.
- **Monochromators**: Used to select specific wavelengths of light for excitation and emission.
- **Detector**: Measures the intensity of the emitted fluorescence.
- **Computer/Recorder**: Analyzes and records the fluorescence data.

6.6.3.2 Applications

The fluorescence spectroscopic technique is invaluable in both research and clinical domains due to its ability to provide detailed information about molecular interactions and dynamics. This spectroscopy is used in the following sectors:

- **Biochemistry**: Studying proteins, nucleic acids, and other biomolecules.
- **Medical diagnostics**: Detecting biomarkers and monitoring diseases.
- **Environmental science**: Analyzing pollutants and toxins.
- **Chemical analysis**: To identify and quantify substances in various samples.

6.6.3.3 Advantages

- **High sensitivity**: Capable of detecting very low concentrations of substances.
- **Specificity**: Can differentiate between closely related compounds.
- **Non-destructive**: Often non-invasive and requires small sample sizes.

6.6.4 Raman Spectroscopy

Raman spectroscopy involves measuring the inelastic scattering of light (Raman scattering) by molecules. Light interacts with a molecule and most of it is elastically scattered (Rayleigh scattering), but a small fraction is scattered at different energies. This change in energy provides information about the vibrational modes of the molecules.

6.6.4.1 Components

- **Laser source**: Provides the monochromatic light (usually visible or near-infrared) to excite the sample.
- **Sample**: The material that is being analyzed.

- **Optical elements**: Lenses and mirrors are used to direct the laser light onto the sample and collect the scattered light.
- **Monochromator/spectrograph**: Separates the Raman scattered light from the Rayleigh scattered light.
- **Detector**: Measures the intensity of the Raman scattered light at different wavelengths.
- **Computer/Recorder**: Analyzes and records the Raman spectrum.

6.6.4.2 Applications

Raman Spectroscopy is widely used in the following:

- **Chemistry**: To identify molecular composition and structure.
- **Materials science**: To analyze the properties of materials, including polymers, nanomaterials, and crystals.
- **Biology**: To study biomolecules like proteins, lipids, and nucleic acids.
- **Pharmaceuticals**: Used for quality control and monitoring drug formulation.
- **Forensics**: to identify substances in forensic investigations.

6.6.4.3 Advantages

- **Non-destructive**: Does not alter or destroy the sample.
- **Minimal sample preparation**: Often requires little to no sample preparation.
- **Versatility**: Can analyze solids, liquids, and gases.

6.6.4.4 Key Features

- **Raman shift**: The difference in energy between the incident and scattered light, providing a unique fingerprint for each molecule.
- **Sensitivity**: Capable of detecting low concentrations of substances.
- **Spatial resolution**: Can be combined with microscopy to provide high-resolution images.

6.7 Other Characterization Techniques

Some other characteristics of the material are also vital to analyze other properties and conditions, such as the thermal behavior of the material, the heat flow during chemical reactions, how phase transition takes place as a function of temperature and time, and environmental conditions. Some characterization techniques are used to analyze the thermal characteristics of the material and working environment, such as TGA, DSC, and a gas analyzer, respectively. Thermal analysis techniques are used to study the properties and behavior of materials as a function of temperature. TGA is used to analyze the change in mass with respect to temperature and time, while DSC analyzes the chemical reactions and phase formation at different temperatures and time conditions. A gas analyzer is used to measure the presence of gas compositions in a particular environment. A detailed explanation is mentioned here for a better understanding of these techniques.

6.7.1 Thermogravimetric Analysis

TGA is a widely used analytical technique that measures the change in a material's mass as a function of temperature or time under a controlled atmosphere. This technique provides valuable insights into the thermal stability, composition, and decomposition characteristics of a material.

6.7.1.1 Components

- **Balance**: A highly sensitive balance to measure the sample's mass accurately.
- **Furnace**: Heats the sample at a controlled rate.
- **Atmosphere control**: Allows for the introduction of different gases (e.g., nitrogen, oxygen) to study the material's behavior under various conditions.
- **Computer/Recorder**: Monitors the mass changes and records the data.

6.7.1.2 Working Principles

- **Sample preparation**: A small sample of the material is placed on the balance inside the furnace.
- **Heating**: The furnace heats the sample at a predetermined rate (e.g., $10°C/min$).
- **Mass measurement**: As the temperature increases, the sample undergoes various thermal events (e.g., evaporation, decomposition), leading to changes in mass.
- **Data analysis**: The balance continuously measures the mass, and the data is plotted as a thermogravimetric curve (mass vs. temperature/time).

6.7.1.3 Applications

TGA is used in various fields, including the following:

- **Polymer analysis**: Studying the thermal stability and composition of polymers.
- **Material science**: Characterizing materials, including metals, ceramics, and composites.
- **Pharmaceuticals**: Analyzing the thermal properties of drugs and excipients.
- **Environmental science**: Determining the composition of complex mixtures, such as soils and waste materials.
- **Quality control**: Ensuring the consistency and purity of materials in manufacturing processes.

6.7.1.4 Advantages

- **High sensitivity**: Capable of detecting small mass changes.
- **Versatility**: Can analyze a wide range of materials.
- **Detailed analysis**: Provides insights into thermal stability and composition.

6.7.2 Differential Scanning Calorimetry

DSC is a sophisticated thermal analysis technique that is used to measure the heat flow associated with phase transitions and chemical reactions in a material as a function of temperature or time. It provides valuable insights into the thermal properties of materials, such as melting points, crystallization, glass transition, and thermal stability.

6.7.2.1 Working Principles

- **Sample preparation**: A small sample of the material is placed in the sample pan, while an inert reference material is placed in the reference pan.
- **Heating/Cooling**: DSC instrument heats or cools both pans at a controlled rate.
- **Heat flow measurement**: As the sample undergoes phase transitions (e.g., melting, crystallization) or chemical reactions, it absorbs or releases heat. The DSC measures the difference in heat flow between the sample and reference pans.
- **Data analysis**: Heat flow data is plotted as a DSC curve (heat flow vs. temperature/time), revealing thermal events in the sample.

6.7.2.2 Components

- **Sample and reference pans**: The sample and an inert reference material are placed in separate pans.
- **Heating unit**: Heats both the sample and reference pans at a controlled rate.
- **Temperature sensors**: Monitor the temperature of both the sample and reference pans.
- **Computer/Recorder**: Records the heat flow and temperature data.

6.7.2.3 Applications

DSC is widely used in various fields, including the following:

- **Polymer analysis**: Studying the thermal properties of polymers, such as melting points, glass transition, and crystallization behavior.
- **Pharmaceuticals**: Analyzing the thermal stability and compatibility of drug formulations.
- **Materials science**: Characterizing the thermal behavior of metals, alloys, ceramics, and composites.
- **Food science**: Investigating the thermal properties of food products, such as fats and oils.
- **Quality control**: Ensuring the consistency and purity of materials in manufacturing processes.

6.7.2.4 Advantages

- **High sensitivity**: Capable of detecting small heat changes.
- **Detailed analysis**: Provides comprehensive information about thermal events.
- **Versatility**: Can analyze a wide range of materials with minimal sample preparation.

6.7.3 Gas Analyzer

A gas analyzer is a device that is used to measure the composition and concentration of gases in a given environment. These instruments are essential in a variety of fields, including environmental monitoring, industrial processes, and safety applications. These analyzers are crucial for maintaining environmental standards, ensuring industrial efficiency, and protecting human health. The types of gas analyzers are mentioned briefly.

6.7.3.1 Types of Gas Analyzers

1. **Portable gas analyzers**: Compact and lightweight, ideal for field-work and on-site measurements.
2. **Fixed gas analyzers**: Permanently installed at a specific location, often used for continuous monitoring in industrial settings.
3. **Single gas analyzers**: Designed to detect and measure one specific gas.
4. **Multi-gas analyzers**: capable of detecting and measuring multiple gases simultaneously.

6.7.3.2 Working Principles

Gas analyzers typically work by using sensors that respond to specific gases. These sensors can be based on various principles, such as the following:

- **Infrared absorption**: Measures the amount of infrared light absorbed by gas.
- **Electrochemical cells**: Generate a current proportional to gas concentration.
- **Catalytic beads**: Detect combustible gases by measuring the heat generated during oxidation.

6.7.3.3 Applications

- **Environmental monitoring**: Track air quality and detect pollutants like CO_2, NOx, and SO_2.
- **Industrial processes**: Ensure optimal combustion, controlling emissions, and monitoring gas purity in chemical production.
- **Safety applications**: Detect hazardous gases in workplaces like mines, factories, and confined spaces to prevent accidents and health risks.

6.7.3.4 Advantages

- **High sensitivity and accuracy**: Ensures precise measurements even at low concentrations.
- **Real-time data**: Provides instant readings and continuous monitoring.
- **Data logging and analysis**: Stores measurement data for later analysis and reporting.
- **User-friendly interface**: Often features easy-to-read displays and intuitive controls.

Theoretical Questions and Problems

1. What is the primary purpose of an advanced characterization technique?

2. What types of information can be obtained using advanced characterization techniques?

3. What are the key factors to consider when choosing an appropriate advanced characterization technique for a specific material analysis?

4. How does the sample preparation process differ for various advanced characterization techniques?

5. What is the fundamental principle of optical microscopy?

6. What is the difference between brightfield and darkfield microscopy?

7. Explain the principle of phase contrast microscopy and when it is useful.

8. What is fluorescence microscopy, and how is it used in advanced characterization?

9. What are the limitations of optical microscopy compared to other advanced characterization techniques like SEM or TEM?

10. What is the main difference between SEM and TEM?

11. What kind of information can be gathered from an SEM image regarding surface morphology and elemental composition?

12. What is a key challenge associated with TEM analysis?

13. What is the principle behind AFM, and what kind of surface details can be observed?

14. How does XRD determine the crystal structure of a material?

15. What is the principle behind X-ray Photoelectron Spectroscopy (XPS), and what information does it provide about surface chemistry?

16. What are the limitations of XRD for analyzing nanostructured materials?

17. How does FTIR spectroscopy identify functional groups within a molecule?

18. What is the main difference between Raman and Infrared spectroscopy?

19. What is the fundamental principle of fluorescence spectroscopy?

20. What are some applications of fluorescence spectroscopy?

21. What is the principle behind DSC, and how is it used to analyze phase transitions?

22. How does TGA measure the weight loss of a material as a function of temperature?

7

Data Management, Optimization, and Post-Processing Techniques

7.1 Introduction

As discussed in Chapter 3, evaluating and assessing the material's response are vital for various applications under different working conditions. Experimental, simulation, and analytical are the three main approaches that could be used to evaluate, analyze, and assessment of the material's response. A large amount of data is generated during the assessment process of any engineering components, and analysis of all of this data is important. The procedure of research data management, various soft computing-based optimization techniques such as GAs, particle swarm optimization (PSO), and ant colony optimization (ACO), could be used for optimizing the research data of any problems you may think of. We will see a detailed description of all these methods in this chapter.

7.2 Research Data Management Process

Data management is an essential part of the research journey because it is needed for analysis and future reference at the different stages of the research project. The basic lifecycle of research data management is presented/mentioned in Figure 7.1. Although it is a separate course for study, some important steps of research data management are discussed briefly below.

7.2.1 Data Organization and Planning

The organization of research data and keeping the relevant references are important to retrieve the data in the future to analyze and kind a reference for other users/researchers. The files should be made/prepared such that they

DOI: 10.1201/9781003664871-7

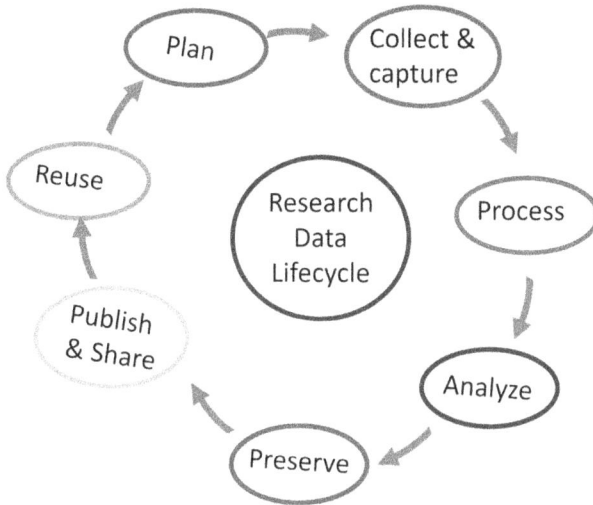

FIGURE 7.1
Schematic representation of the research data management life cycle.

are self-explanatory. Data could be in different forms, such as histograms, pie charts, bar charts, images, videos, and audio; thus, they should be kept in different folders with clear descriptions about them. Also, the relevant details (experimental details, type of material, initial conditions, and so on) should be provided in a separate Word file.

7.2.2 Data Storage and Backup

The data should be stored in different places, and there must be a backup of each file with all stakeholders. Data storage and backup are essential for a researcher and a businessman working with digital information. It is important to protect the research data against accidents, cyberattacks, human error, and natural disasters. You may store data in cloud space or other online platforms to avoid the aforementioned problems, but make sure that research data is protected from any kind of cyber-attack and loss.

7.2.3 Data Cleaning and Quality Control

While capturing/acquiring the data, the quality must be maintained (as per the guidelines mentioned for the particular experiment) so that it can be stored and used for a long time. The cleaning of data could be needed for some time to remove the unwanted noise/interruptions and do further analysis. There are sets of rules and protocols that should be followed during the data cleaning and quality control of the researcher's data.

7.2.4 Data Sharing and Privacy

The data needs to be shared with the publishers/co-workers/institutions to validate your findings and research outcomes. Thus, it should be shared through proper channels so that you can keep a record of it, and if needed, you can produce/claim it. Also, make sure that the privacy of data is maintained by all the stakeholders during the different stages of the research journey.

7.3 Soft Computing Optimization Techniques

7.3.1 Genetic Algorithms

This is based on the evolution theory, which considers the survival of the fittest. In GAs, we will start with an initial population (each individual is assigned a fitness value) and select parents that could undergo recombination and mutation (like in natural genetics), producing new children, and the process is repeated over various generations. In these processes, individuals are given a higher chance to mate and yield more "fitter" individuals to get the best offspring, as schematically shown in Figure 7.2. GAs are

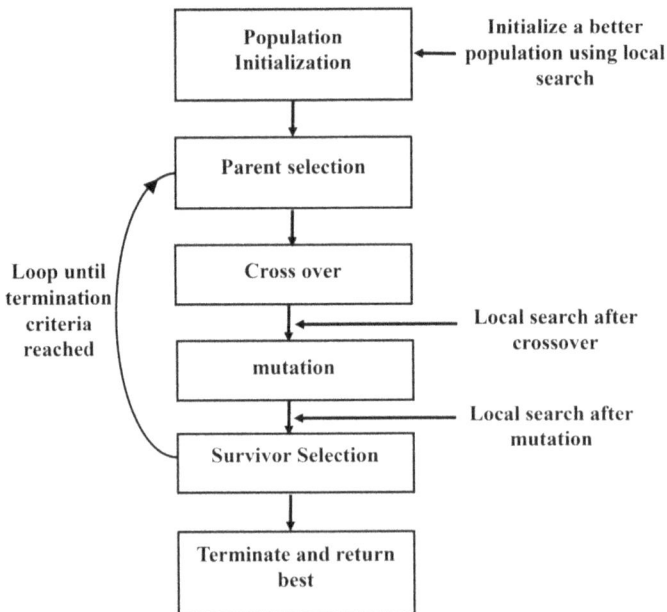

FIGURE 7.2
The basic structure of the GA.

sufficiently randomized in nature, but they perform much better than a random local search.

Advantages of GAs:
- It is faster and more efficient compared to other methods.
- It gives several possible solutions and not just a single one. It could be used for any complex problem to get solutions that get better over time.
- It could be more useful when there are a large number of parameters associated with the problem.
- Multi-objective problems, including continuous and discrete, could also be optimized.

Limitations of GAs:
- The use of GAs is not suitable for simple problems due to the computationally expensive process.
- More iterations (hence more time) for estimating the fitness value might be needed for some problems.
- There are no guarantees on the quality of the solution.

7.3.2 Particle Swarm Optimization

PSO is a powerful optimization algorithm inspired by the swarm behavior observed in nature, such as fish and bird schooling. It is different from other optimization algorithms in such a way that only the objective function is needed, and it is not dependent on the gradient or any differential form of the objective. Multiple particles (each particle called a candidate) form the swarm, and they co-exist and corporate simultaneously. Each particle looks for the best solution by an iterative approach and searches all the areas. Furthermore, each particle keeps track of its personal best solution (optimum), as well as the best solution (optimum) in the swarm. Based on the personal best and the global best solution, each particle dynamically adjusts its parameters to get the optimized solution to the problem. You may take the example of the position and velocity of the individual bird in the group for optimization using the PSO, concerning the average velocity and position of the group approach. A flowchart diagram of PSO techniques is given in Figure 7.3.

Advantages of PSO:
1. Fewer parameters are needed to perform PSO.
2. Efficient global search algorithm compared to others.
3. Insensitive to scaling of design variables.

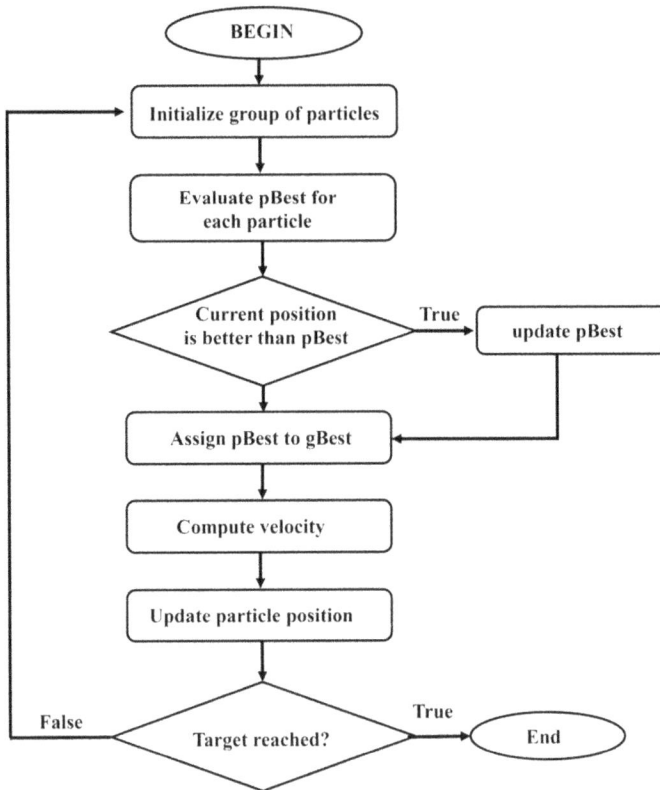

FIGURE 7.3
Flow diagram of the PSO.

7.3.3 Ant Colony Optimization

The principle of ant (live-in community nests) colony optimization is to observe the movement of the ants from their nests to search for food in the shortest possible path. It is observed that, initially, ants move randomly surrounding the nest, thus leading to multiple paths from the nest to the food source. Generally, the pheromone concentration increases when ants return to nests with food on a particular path, and then a greater number of the ants follow (depending on the concentration and the rate of evaporation of pheromone) the same path to the food location. ACO algorithms could be understood by the process given in Figure 7.4. The steps of the optimization techniques are as follows:

- **Stage 1**: Initially, there is no pheromone content on the paths; thus, the possibility of selecting the path is equal (0.33). Ants can take any path randomly.

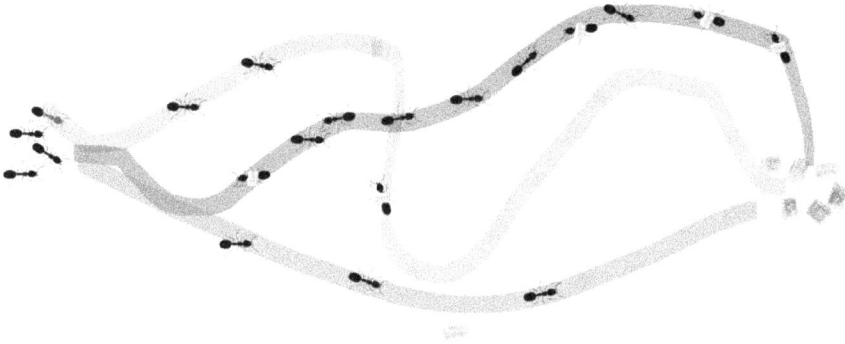

FIGURE 7.4
ACO technique.

- **Stage 2**: Ants start to search along these three paths, and if the speed of all ants same, then the time for reaching the food location would be less on the shortest path.
- **Stage 3**: Once they took the food then they would take the same path for the return journey due to the higher concentration of phero-mone along the shorter path already available, thus, the probability of selection is higher.
- **Stage 4**: More ants return via the shorter path (due to evaporation, the pheromone concentration in the longer path reduces), and subse-quently, the pheromone concentrations also increase. Therefore, the entire colony gradually adopts the shorter path with a higher prob-ability. So, path optimization is attained.

7.4 Data Processing Techniques

7.4.1 Introduction to Python

It allows programmers to write in fewer lines, and thus, you can work quickly and integrate systems more efficiently. The following are the primary features of the Python language:

- Interpreted: No need for separate compilation and execution
- Platform Independent: It can be used on Linux, Windows, and many other platforms.wp
- Free and Open Source: It is openly available.

- High-level Language: No need to worry about the program size.
- Simple and Robust: Easy to learn and can handle difficult problems.
- Rich Library Support: The standard library is vast.

Python program using variables

myNumber = 7

print(myNumber)

myNumber2 = "Hello world"

print(myNumber2)

Output of program

7

Hello world

Pros:

1. Could make a simple code with ease of writing.
2. Programmers can express themselves in fewer lines of code compared to other languages.
3. Python supports multiple programming paradigms, like object-oriented and imperative.
4. The standard inbuilt library is vast.

Cons:

1. Slow speed of execution compared to C++.
2. Absence from mobile computing and browsers.

7.4.2 Introduction to MATLAB

Matrix Laboratory (MATLAB) is a high-performance language used for writing code to solve a particular problem. It is primarily used to solve mathematical equations with high precision in less time. MATLAB software allows us to do matrix manipulations, plotting of mathematical complicated/complex functions, various algorithms implementation, and creation of user interfaces to develop our code. The main advantage of the MATLAB library comes with a set of many inbuilt functions that could be used to perform various operations like finding the inverse and determinant of a matrix, cross product, and dot product. Moreover, no header files (like C, C++) need to be initialized in the beginning for declaring a variable and its data type. User-friendly interface and an inbuilt library of MATLAB make it the best language for doing mathematical operations efficiently.

MATLAB Software Interface

There are several dialogue boxes used for various purposes. Some of the important ones are as follows:

- **Command window**: This is the main window used to type and execute statements quickly. It is generally used for small, easily executable programs.
- **Editor (Script)**: This is used to write and execute larger programs with multiple statements and complex functions. The file can be saved for future operations with the file extension ".m".
- **Workspace**: In the command window, the values assigned to the variables during the program (in the editor) are displayed in the workspace.
- **Command history window**: This window displays a log of statements that you ran in the current and previous MATLAB sessions. The command history lists the time and date of each session in the short date format for your operating system.

7.4.3 Introduction to Data Visualization Tools

There are several tools/software available to analyze and visualize the research data. One of them is the Origin software, which is widely used by engineers/scientists; it is a powerful data analysis and publication-quality graphing software, tailored to the needs of scientists and engineers. Users can customize and automate the research data, do the required analysis, make the plots/graphs, and do reporting tasks. Users can make a "template" for later use based on their requirements. For making the graphs, doing the fitting, applying various statistical conditions, and finishing the visualization of the plots. There is a dedicated website for knowing more about its functioning and capabilities; users can refer to that for more details.

Theoretical Questions and Problems

1. Why is the research data management an important step in the research journey?
2. Could you write the various steps of research data management? Explain them briefly.

3. Do we need the optimization techniques for analyzing the research data? How can we improve the efficiency of the process?

4. What are various soft computing-based methods for optimizing the research data?

5. What is the principle of the GA optimization technique? Explain the method in detail, with advantages and limitations.

6. What is the principle of the PSO technique? Explain the method in detail, with advantages and limitations.

7. What is the principle of the ACO technique? Explain the method in detail, with advantages and limitations.

8. What are various data-processing techniques for visualizing the research results/outcomes?

9. Why is Python popular for data processing? Explain the method in detail, with advantages and limitations.

10. What is the difference between Python and MATLAB? Which one is best for data processing and why?

11. Why is MATLAB a high-performance language? Explain the method in detail, with advantages and limitations.

12. What is the difference between high and low-performance language? Could you mention the names of the few low-performance languages?

13. Why is the proper representation of data vital for publication in high-impact journals?

8

Documentation and Art of Scientific Writing

8.1 Need for Scientific Writing

Scientific writing is essential for effectively communicating research findings and advancing knowledge across various disciplines. It provides a structured and precise way to present complex ideas, ensuring clarity, accuracy, and reproducibility. Unlike casual or literary writing, scientific writing follows a formal and objective tone, allowing researchers to convey their work in a logical and standardized manner. This is crucial for peer review, collaboration, and the accumulation of knowledge, as it enables others to validate, build upon, or challenge existing research. Additionally, scientific writing plays a vital role in bridging the gap between researchers and the broader community, including policymakers and industry professionals, by translating technical findings into actionable insights. Without clear and well-organized scientific writing, valuable research could be misinterpreted, overlooked, or rendered inaccessible, which leads to hindering progress and innovation. Below are key reasons why scientific writing is important, as shown in Figure 8.1.

8.1.1 Knowledge Dissemination

Scientific writing allows researchers to share their findings with specific audiences in a structured and understandable manner to enhance understanding, awareness, and application. This dissemination contributes to the collective pool of knowledge and helps advance science. It plays a crucial role in education, research, and professional development by ensuring that valuable insights and innovations reach the right people. Knowledge can be disseminated through various channels, including academic publications, conferences, workshops, digital platforms, and informal discussions. The effectiveness of dissemination depends on factors such as the clarity of communication, accessibility of information, and engagement with the audience.

DOI: 10.1201/9781003664871-8

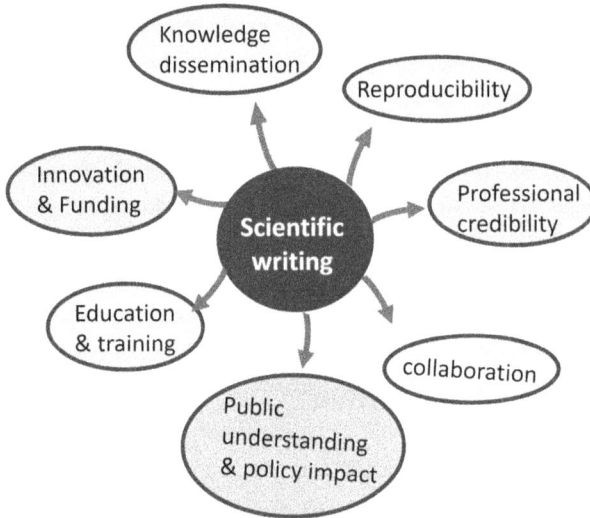

FIGURE 8.1
Schematic of scientific writing.

8.1.2 Reproducibility

Clear and precise writing enables other researchers to replicate experiments and validate results. It involves providing clear, detailed, and transparent descriptions of methodologies, data, and analyses so that others can achieve the same results under similar conditions. Reproducibility enhances the credibility of scientific work, prevents errors, and fosters trust within the research community. Key elements include proper documentation of experimental procedures, sharing raw data, and using standardized reporting formats. As science advances, reproducibility remains essential for validating discoveries, guiding future studies, and maintaining the integrity of academic and industrial research. This reproducibility is crucial for verifying scientific discoveries and maintaining the integrity of research.

8.1.3 Professional Credibility

High-quality scientific writing reflects the rigor and professionalism of the researcher. Authors enhance their credibility by providing well-researched, evidence-based findings, properly citing sources, and ensuring reproducibility in their work. Clear and objective communication, free from bias or exaggeration, further strengthens professional integrity. Peer-reviewed publications, ethical authorship practices, and openness to constructive criticism also contribute to a researcher's credibility. It builds trust in their work and enhances their reputation in the scientific community.

8.1.4 Collaboration Opportunities

Scientific papers and reports serve as a basis for collaboration. It is a vital aspect of research that brings together experts from diverse fields to share knowledge, resources, and skills in pursuit of common scientific goals. Collaborative efforts enhance innovation by combining different perspectives, methodologies, and technologies, leading to more comprehensive and impactful discoveries. Effective collaboration nurtures open communication, ethical research practices, and the sharing of data and findings, ultimately improving the quality and credibility of scientific work. Well-documented research can inspire partnerships and encourage interdisciplinary projects.

8.1.5 Public Understanding and Policy Impact

Science writing tailored for a general audience plays a crucial role in making complex research accessible and engaging, helping to bridge the gap between scientists and the public. Additionally, well-documented research serves as a valuable resource for policymakers, enabling them to make evidence-based decisions that address general challenges effectively. When scientific findings are communicated clearly and accurately, they contribute to more informed policy development, ensuring that decisions are grounded in reliable data rather than misinformation or speculation. Thus, effective science writing benefits both society and decision-makers by promoting transparency, trust, and progress in various fields.

8.1.6 Permanent Record

Publications create a permanent record of scientific work. They ensure that knowledge persists over time and can be revisited for future studies or applications. Scientific writing serves as a permanent record of scientific discoveries, experiments, and theories. It allows for the dissemination of knowledge, collaboration among researchers, and the building upon previous work.

8.1.7 Education and Training

Scientific literature serves as an educational resource for students, educators, and early-career researchers, promoting learning and skill development in the field. It also helps in foundational knowledge for scientific methods by understanding the principles of hypothesis formulation, experimental design, data collection, analysis, and interpretation. It also imparts domain-specific knowledge and a deep understanding of the chosen field of research (e.g., biology, physics, social sciences).

8.1.8 Innovation and Funding

Well-articulated research proposals and papers are vital for securing funding and inspiring innovation. They demonstrate the feasibility and significance of the work to potential sponsors.

In essence, scientific writing is a cornerstone of the scientific method, enabling the clear, precise, and impactful exchange of ideas and discoveries.

8.2 Selection of a Suitable Journal for Publication

Selecting the right journal for publication is a critical step in broadcasting your research effectively, ensuring it reaches the appropriate audience and achieves the intended purpose. A well-matched journal enhances the visibility, credibility, and influence of your work within the academic and professional community. To begin, identify the scope and focus of your research and compare it with potential journals to determine the best fit. Consider the target audience, whether it consists of researchers, practitioners, or policymakers, to maximize engagement. Evaluate the journal's impact factor, indexing in major databases, and reputation within your field, as these factors influence the visibility and recognition of your work. Review past publications to assess the relevance of your topic and ensure alignment with the journal's thematic preferences. Additionally, consider the journal's publication frequency, acceptance rate, and review process to gauge the likelihood of acceptance and the timeline for publication. Open-access options can also be beneficial if you want to ensure broader accessibility to your research. Finally, carefully read the author guidelines to adhere to submission requirements and formatting standards, increasing the chances of a smooth review process. By following these steps, you can strategically select a journal that enhances your research's impact and contributes meaningfully to your academic or professional field and has been represented in Figure 8.2.

1. **Define your research goals**
   a. Identify the purpose of your publication, whether it is materials science, aerospace engineering, polymer technology, or high-temperature materials. Is it to share novel findings, gain recognition, or influence a particular field?
   b. Determine the primary audience for your research: experts, interdisciplinary readers, or policymakers. Choose a journal that specializes in your field to ensure that your work aligns with its scope.

Selection of a proper journal

Evaluating the scope of journal

Assess the journals reputation

Check for peer reviewed status

look at indexing and achieving

Avoid predatory journals

Match formatting and submission requirements

Check ethical standard

Consult colleague and supervisor

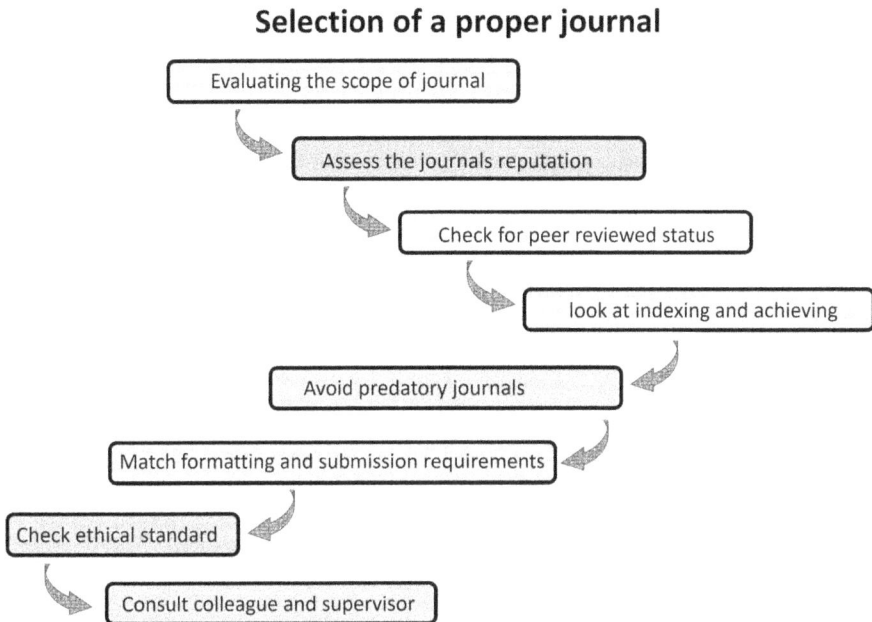

FIGURE 8.2
Why selection of a scientific journal is important.

2. **Evaluate the scope of the journal**
   a. Review the journal's aims and scope to ensure your research fits their thematic focus.
   b. Look at recent issues to see if similar studies have been published.

3. **Assess the journal's reputation**
   a. Check the impact factor or other bibliometric indicators to evaluate its reputation and influence in the field.
   b. Consider the journal's ranking in your discipline (e.g., quartiles in Scopus or Web of Science).

4. **Consider open access versus subscription-based journals**
   a. Decide whether you prefer an open-access journal for broader accessibility or a traditional journal for a more focused academic audience.
   b. Be mindful of publication fees, such as article processing charges (APCs), in open-access journals.
   c. Each journal has specific formatting, word count, and citation requirements. Make sure your manuscript fits their submission criteria to avoid rejection due to formatting issues.

5. **Check for peer-reviewed status**

   a. Ensure the journal has a rigorous peer-review process to validate your research and enhance its credibility.

   b. Also, consider publication timelines; some journals take months to review and publish, while others offer faster turnaround times.

6. **Review journal metrics**

   a. Analyze acceptance rates, average time to first decision, and publication timelines.

   b. Some journals offer a fast-track review option, which may be useful if you need quick publication.

7. **Look at indexing and archiving**

   a. Ensure the journal is indexed in major databases like PubMed, Scopus, Web of Science, or IEEE Xplore to maximize visibility depending on your domain. Open-access journals provide greater reach but may require publication fees.

8. **Avoid predatory journals**

   a. Beware of journals that lack peer review, have excessive publication fees, or solicit manuscripts aggressively.

   b. Verify journals through trusted sources like DOAJ (Directory of Open Access Journals), COPE (Committee on Publication Ethics), or publisher websites.

9. **Consult colleagues and supervisors**

   a. Seek advice from experienced researchers, mentors, or colleagues who have published in your field.

   b. They may recommend journals with a good track record for similar research.

10. **Match formatting and submission requirements**

    a. Review the journal's author guidelines for formatting, word limits, and reference styles.

    b. Choose a journal that aligns with the structure and style of your manuscript.

11. **Use journal matching tools**

    a. Tools like Elsevier Journal Finder, Springer Journal Suggester, or EndNote manuscript matcher can suggest journals based on your title, abstract, and keywords.

12. **Check ethical standards**
   a. Verify that the journal adheres to ethical guidelines (e.g., COPE – Committee on Publication Ethics or ICMJE – International Committee of Medical Journal Editors standards) for research integrity and publication practices.

8.3 Structure of Manuscript

A well-structured manuscript follows a standardized format to ensure clarity, coherence, and effective communication of research findings. The common structure includes the following key sections and has been represented by a flow diagram as shown in Figure 8.3.

1. **Title page**
   The title should summarize the research in a single, compelling sentence. The title should convey the main focus of the research while

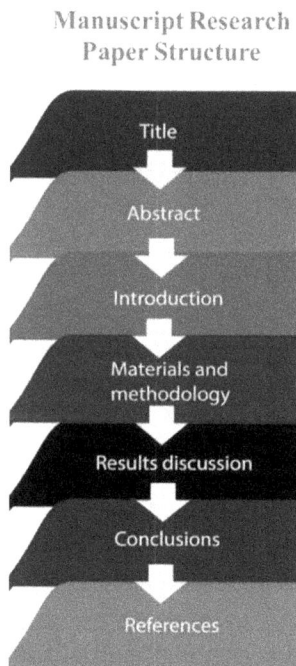

FIGURE 8.3
Flow diagram for writing the structure of a journal manuscript.

being as brief as possible. Avoid unnecessary jargon, abbreviations, or complex terms that may confuse readers. It should accurately reflect the content, methodology, and findings of the study to ensure that readers and researchers searching for related work can easily identify its relevance.

Author Information should include the full names of all authors. Their state affiliations should also be mentioned, using superscript numbers if necessary. The corresponding Author indicates one person responsible for correspondence (with email and phone number).

2. **Abstract**

An abstract in a journal article is a concise summary of the research, providing readers with a quick overview of the study's key aspects. Abstract should be defined with the background of why the study was conducted. It should clearly define the objective of the study with defined goals and a hypothesis. It should include methods like key techniques and approaches. It includes the main results, principal findings, and highlights key statistics. At the end, it should deliver the conclusion and implications of the study. Write the abstract in the past tense for methods and results. Be precise but concise (150–300 words, as per journal rules). Avoid references or overly technical terms.

For keywords, select five to seven keywords relevant to your study to enhance discoverability. When readers or researchers search for information in academic databases, search engines, or journal repositories, keywords help index and categorize articles, making them easier to find. Well-chosen keywords increase the likelihood of an article appearing in relevant search results, thereby expanding its reach and potential impact.

3. **Introduction**

In the context part, start the introduction by explaining the background or global relevance of the research. It should contain the previous relevant literature survey done in a similar domain to define the research gaps and objectives of the current manuscript. It should clearly define the research gaps. It should identify a specific problem or gap in existing literature. The introduction should clearly state your objectives and hypotheses in the current analysis.

4. **Methodology (materials and experimental methods)**

The methodology section explains the overall research approach and rationale behind the chosen methods. It provides a theoretical framework, discussing why a specific research design (e.g., qualitative, quantitative, or mixed methods) was selected. This section also outlines the philosophical or scientific principles guiding the study, such as experimental, observational, or case study approaches. The methods section provides a step-by-step description of the procedures used in the research. It details data collection techniques (e.g., surveys, experiments, interviews), sample selection criteria, materials or equipment used, and data analysis techniques. The goal of this section is to provide enough detail for other researchers to replicate the study if needed.

5. **Results**

An organized presentation of research findings is essential for clarity, coherence, and effective communication. To maintain logical flow, results should be presented in the same order as the methods section to ensure that readers can easily follow the progression of the study. This structured approach helps maintain consistency and prevents confusion. Objective phrasing ensures that conclusions are based on empirical evidence rather than personal interpretation. Phrases such as "As shown in Table 1" or "Figure 2 illustrates" help guide readers to specific data points, reinforcing transparency and credibility. Visual aids, including graphs, tables, and charts, play a crucial role in presenting complex data effectively. These elements should be well-labeled, with clear legends, units, and axis titles to enhance understanding. Captions should be descriptive enough to stand alone, allowing readers to grasp the key message without referring back to the main text. A well-designed figure or table can summarize findings more efficiently than lengthy descriptions, making the results more accessible.

6. **Discussion**

The discussion section in a manuscript serves as a critical platform for researchers to interpret their findings and contextualize them within the broader landscape of existing literature. Unlike the results section, which presents data in a straightforward manner, the discussion offers a nuanced analysis that goes beyond mere restatement. Here, researchers delve into the implications of their findings, exploring what the results indicate about the research question and how they align with or challenge previous studies. This section not only highlights the significance of the research within the academic community but also emphasizes its potential impact on practical applications, policy, or future innovations. Furthermore, the discussion allows authors to address any limitations of their study, fostering transparency and encouraging readers to consider the broader applicability of the results. Ultimately, by connecting their work to ongoing scholarly conversations and proposing avenues for future research, authors enhance the value of their contributions and underscore the importance of their findings in advancing knowledge within the field.

7. **Conclusion**

The Conclusion section in a manuscript is the final part where the key findings of the study are summarized, their broader implications are highlighted, and potential future research directions are suggested. Unlike the discussion section, which provides a detailed interpretation and comparison with previous studies, the conclusion focuses on reinforcing the study's main message concisely and effectively.

Purpose of the conclusion section:

a. Reinforce the study's main takeaways.

b. Highlight its contribution to existing knowledge.

c. Provide a clear, final statement on the study's impact.

d. Inspire further research and applications.

e. A well-written conclusion leaves readers with a strong understanding of the study's importance and a clear direction for future investigations.

8. **Funding and acknowledgments**

The funding and acknowledgments section of a research article serves an essential role in recognizing the financial and institutional support that made the study possible. This section typically includes details about grants, sponsorships, or funding agencies that contributed resources to the research, ensuring transparency and credibility. Acknowledgments also extend to individuals, colleagues, or institutions that provided assistance, such as technical support, data analysis, or critical feedback, but who do not meet the criteria for authorship. By explicitly mentioning funding sources and contributors, researchers uphold ethical standards and demonstrate accountability, which is crucial for maintaining trust in scientific work. Furthermore, proper acknowledgment of financial backing allows funding bodies to track the impact of their investments and encourages continued support for future research endeavors. This section, though often brief, is a crucial part of scholarly communication, reflecting the collaborative nature of scientific discovery.

9. **References**

The References section in a manuscript is a comprehensive list of all the sources cited throughout the paper. It provides full bibliographic details of books, journal articles, conference papers, reports, and other materials referenced in the study. This section ensures that credit is given to original authors, supports the credibility of the research, and allows readers to locate the cited works for further study.

Key features of the reference section

a. **Accuracy and consistency** – Citations must be correctly formatted according to the journal's required referencing style, such as APA, MLA, Chicago, or IEEE. Inaccurate or incomplete references can lead to manuscript rejection.

b. **Relevance** – Only include sources directly cited in the text. References should be up-to-date and relevant to the research topic.

c. **Alphabetical or numerical order** – Depending on the chosen citation style, references are listed either alphabetically (e.g., APA, Chicago) or numerically (e.g., IEEE, Vancouver) based on the order of citation in the text.

d. **DOI or URLs for online sources** – Whenever possible, include DOI (Digital Object Identifier) or stable URLs for journal articles to ensure easy access.

e. **Facilitates verification** – By allowing readers to find the cited studies.

A well-structured references section ensures academic integrity and strengthens the manuscript's scholarly impact.

8.4 Abstract

An abstract is a brief summary of a research study, highlighting its objective, methods, key findings, and conclusions. It helps readers quickly assess the study's relevance. A well-crafted abstract is clear, informative, and self-contained, allowing understanding without reading the full paper. As the first section encountered by readers and databases, it plays a crucial role in academic publishing. Writing a concise and effective abstract for a journal involves summarizing the key points of your paper in a clear and concise manner. The flow diagram for writing a standard abstract is shown in Figure 8.4, and here's how to do it:

1. **Purpose** – Art with the reason for conducting the study. Briefly explain the problem or research question your work addresses. The abstract gives readers a quick overview of the paper's main objectives, methods, findings, and conclusions without requiring them

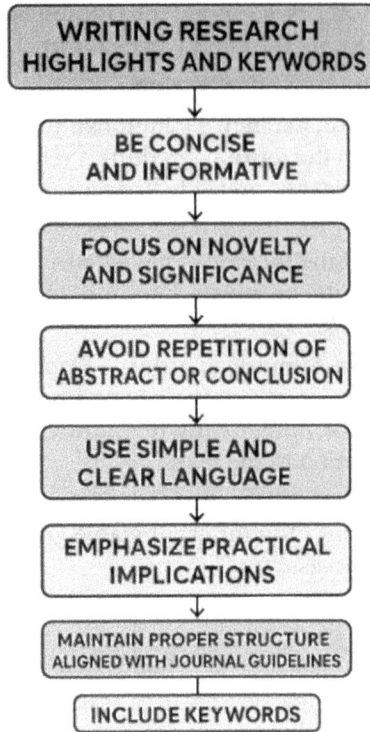

FIGURE 8.4
Flow diagram to show how to write a standard abstract for a manuscript.

to read the full text. A well-written abstract grabs the attention of readers, helping them decide whether the paper is relevant to their interests or field of study. It allows researchers, students, and professionals to understand the essence of your work at a glance, saving them time.

2. **Methods** – Briefly describe the methods or approach you used to conduct your research. Include important techniques, tools, or frameworks. It is an essential component because it explains how you arrived at your findings, giving credibility to your results.

3. **Results** – The results section of the abstract summarizes the key findings of your research. It highlights the most important outcomes and demonstrates how they address the research objectives. The results should be concise, specific, and written in a way that engages the reader without delving into excessive detail or analysis. This focuses on the most significant outcomes relevant to your research objectives.

4. **Conclusions** – State the implications of your findings and their contribution to the field.

   Example: "These findings highlight the critical role of urban green spaces in mitigating pollution and emphasize the need for sustainable city planning."

5. **Keep it brief** – The word limit typically lies between range of 150–250 words (follow journal guidelines). Avoid excessive jargon, detailed data, and citations. Always use the active voice as it makes the abstract more engaging.

8.5 Highlights and Keywords

The highlights and keywords sections of a manuscript play a crucial role in enhancing its visibility, accessibility, and impact. Highlights provide a concise summary of the study's key contributions, typically in bullet points, helping readers quickly grasp the most significant aspects of the research. They should emphasize the novel findings, practical applications, and broader implications of the study. Well-crafted highlights attract attention and can increase the chances of the article being read and cited.

8.5.1 Important Considerations and Points for Writing Research Highlights

- **Concise and informative** – Research highlights should be brief yet impactful, typically consisting of three to five bullet points, each containing 85 characters or fewer. They should effectively summarize the key contributions of the study without unnecessary details.

- **Focus on novelty and significance** – Clearly state what makes the research unique and how it advances existing knowledge. Highlight any breakthroughs, novel methodologies, or important discoveries.

- **Avoid repetition of abstract or conclusion** – Research highlights should not be copied directly from the abstract. Instead, they should distill the most important aspects of the study into short, standalone points.

- **Use simple and clear language** – The highlights should be easily understandable to a broad audience, including researchers from different fields. Avoid jargon and overly technical terms unless necessary.

- **Emphasize practical implications** – If the research has real-world applications (e.g., policy changes, technological advancements, or medical breakthroughs), mention them concisely.

- **Maintain proper structure** – Each bullet point should be written in an active voice and should focus on one key aspect of the study.
- **Align with Journal Guidelines** – Some journals have specific requirements for research highlights, such as word count or formatting rules. Always check and follow these guidelines before submission.

By crafting clear, concise, and compelling research highlights, authors can effectively showcase the importance of their study and enhance its visibility in academic and professional circles.

8.5.2 Keywords

Keywords serve as essential indexing tools that improve searchability in academic databases and search engines. Carefully selected keywords should reflect the core concepts, methodologies, and subject areas of the research, ensuring that the article reaches the right audience. A combination of general and specific terms including standardized scientific terms, enhances discoverability. By effectively summarizing the study in the highlights and optimizing the selection of keywords, researchers can maximize the impact and reach of their work within the academic community.

8.6 Introduction

The introduction section of a journal article provides the background and context for the research, helping readers to understand its significance. It typically begins with a broad discussion of the topic, highlighting existing knowledge, key challenges, or research gaps in the targeted field. After that, the introduction narrows down to the specific research problem, explaining why the study is necessary and how it addresses the identified research gaps. It also includes the research objectives, hypotheses (if applicable), and a brief overview of the study's approach. To write an effective introduction, researchers should ensure clarity, logical flow, and engagement, avoiding unnecessary details while maintaining a strong connection to existing literature. A well-structured introduction sets the stage for the rest of the paper, guiding readers toward the study's core purpose and expected contributions.

8.6.1 How to Write an Effective Introduction?

8.6.1.1 Background and Context

Start with a broad overview of the research topic, explaining its importance and relevance in the field. Use references to previous studies to establish existing knowledge. The background and context section of the introduction in a manuscript provides essential information to help readers understand the broader relevance of the study. It begins by introducing the research topic, explaining its significance in the field, and highlighting key developments or existing knowledge. This section often includes references to previous studies to establish what is already known and identify gaps or limitations in current research. By presenting the historical and theoretical context, it sets the stage for the research problem and justifies the need for the study. A well-written background and context section ensures clarity, engages the reader, and provides a strong foundation for the research objectives and hypotheses that follow.

8.6.1.2 Problem Statement

Identify the gap in research or the specific problem your study aims to solve. Explain why addressing this issue is crucial. The problem statement in the introduction section of a manuscript clearly defines the specific issue or gap in knowledge that the study aims to address. It builds upon the background and context by identifying limitations in existing research, unresolved questions, or practical challenges that require further investigation. A well-articulated problem statement highlights why the issue is important and how it impacts the field, society, or a particular industry. It should be precise, focused, and supported by relevant literature to demonstrate its significance. By clearly stating the problem, this section helps justify the need for the study and guides the development of research objectives and hypotheses, ensuring a logical flow throughout the manuscript.

8.6.1.3 Objectives and Research Questions

Clearly state the purpose of the study, the research questions, or hypotheses that guide the investigation. This section of the introduction in a manuscript outlines the specific goals of the study and the key questions it seeks to answer. Research objectives provide a clear direction for the study, defining what the researchers aim to achieve, whether it is to explore a phenomenon, test a hypothesis, or develop a new methodology. These objectives should be precise, measurable, and aligned with the problem statement. Research questions, on the other hand, break down the study's purpose into specific inquiries that guide data collection and analysis. Well-formulated research questions ensure clarity and focus, helping readers understand the study's scope and expected contributions. Clearly defining objectives and research questions strengthens the study's foundation, demonstrating its relevance and guiding the overall research process.

8.6.1.4 Significance of the Study

Explain the potential impact of your research and how it contributes to the field, whether through new insights, practical applications, or theoretical advancements. The significance of the study section in the introduction of a manuscript explains the importance and potential impact of the research. It highlights how the study contributes to existing knowledge, fills research gaps, or offers practical applications in a specific field. This section may discuss theoretical advancements, policy implications, or benefits for industries, organizations, or society as a whole. Demonstrating why the research matters helps justify the study and encourages further exploration. A well-written significance section strengthens the manuscript by showing its relevance, originality, and potential to drive meaningful change or innovation.

8.6.1.5 Structure of the Paper

In some cases, authors briefly outline the structure of the paper to help readers navigate the content. A well-structured introduction ensures clarity and engages readers, providing them with a strong understanding of the study's purpose and relevance.

8.7 Methodology

The methodology and methods sections of a manuscript provide a detailed explanation of how the research was conducted, ensuring transparency and reproducibility. These sections allow other researchers to evaluate the study's reliability and replicate the experiment or analysis if needed. The methodology explains the theoretical framework and research design. This section describes the specific procedures used in the study. Below is a structured approach to writing these sections in a manuscript.

8.7.1 Key Elements of the Methods Section

1. **Research design** – This describes the overall structure and framework of the study, such as whether it follows a qualitative, quantitative,

mixed-methods, experimental, or observational approach. This design choice dictates how data is gathered and analyzed.

2. **Participants/Sample** – This describes to study population, including sample size, selection criteria, recruitment strategies, and demographic characteristics. This section ensures clarity on who was involved in the study and how they were chosen, which is crucial for assessing generalizability.

3. **Data collection methods** – section explains the techniques used to gather information, such as surveys, interviews, focus groups, laboratory experiments, or secondary data sources. The reliability and validity of these methods significantly impact the study's credibility. It provides details about the tools, instruments, or software used (e.g., questionnaires, laboratory equipment).

4. **Procedures** – This section provides a step-by-step explanation of how the research was carried out, including any interventions, ethical considerations, and protocols followed to ensure consistency.

5. **Data analysis** – This section describes how the collected data was processed and interpreted, including statistical methods, coding techniques, software used, and any models applied. A well-structured methods section enhances the study's transparency and allows other researchers to replicate or build upon the findings, reinforcing the validity of the research.

8.8 Results

The results section of a research manuscript serves as the core presentation of the study's findings, offering a clear and objective report of the data collected. This section is crucial as it provides the factual evidence that supports the study's conclusions. It should be structured logically, following the sequence of research questions or hypotheses, ensuring that readers can easily follow the flow of information. The presentation should be precise, avoiding any interpretation or discussion, which is reserved for later sections. To enhance clarity, the results should be supported by relevant data, including numerical findings, statistical analyses, and qualitative observations if applicable. The use of tables, figures, and graphs is essential for visually representing complex data, making it easier for readers to comprehend patterns and trends. Proper labeling of these visuals, along with concise explanations, ensures that the information is well-organized and accessible. Additionally, findings should be reported with appropriate statistical measures, such as means, standard deviations, and confidence intervals, to establish significance and reliability. Some points to be considered before writing the results of manuscripts are.

1. **Organize results logically** – Present the findings in a logical sequence that aligns with the research objectives or hypotheses, ensuring a clear and structured flow of information. If the results cover multiple aspects or variables, use appropriate headings or subheadings to enhance readability. Begin with the primary findings that directly address the main research questions before moving on to secondary or exploratory results, maintaining coherence and clarity throughout the section.

2. **Report data clearly** – Present numerical or qualitative data concisely and precisely, ensuring clarity and accuracy. Use appropriate statistical measures such as means, percentages, confidence intervals, and p-values to support the findings. Avoid redundant repetition of data in the text if it is already displayed in tables or figures, keeping the section clear and focused.

3. **Use tables, figures, and graphs effectively** – Use charts, graphs, and tables to visually present complex data for better clarity and comprehension. Ensure all visuals are clearly labeled with appropriate titles, legends, and units of measurement, making them easy to interpret.

4. **Report statistical results** – Present descriptive statistics such as mean and standard deviation to summarize the data effectively. If hypothesis testing is involved, include inferential statistics like ANOVA or regression analysis to highlight significant relationships and differences within the data.

5. **Maintain Clarity and Objectivity** – Use concise and precise language, avoiding vague statements to maintain clarity. Present only relevant results without unnecessary details, ensuring the focus remains on key findings. Maintain consistency in data reporting, such as using uniform decimal places, to enhance accuracy and readability.

8.9 Discussion

The discussion section of a manuscript interprets and explains the significance of the research findings about the study's objectives, existing literature, and broader implications. Unlike the results section, which presents data objectively, the discussion provides an analysis of what the findings mean, how they compare with previous research, and their potential impact. This section often includes an evaluation of the study's strengths and limitations, addressing any unexpected results or biases. Additionally, it may suggest practical applications, theoretical contributions, and directions for future research. A well-structured discussion connects the findings to the bigger picture, reinforcing the study's relevance and importance.

8.9.1 Key Elements of the Discussion Section

Summary of Key Findings: Start by briefly reiterating the main results without repeating data. Highlight the most significant trends, patterns, or unexpected observations.

Comparison with previous studies – Relate your findings to existing literature, showing how they support, contradict, or expand upon previous work. This contextualization helps establish credibility and situates the study within the broader scientific conversation.

Interpretation and implications – Explain what the results mean in a broader context. Discuss their potential impact on the field, practical applications, policy implications, or how they advance theoretical understanding. Explain why the results for a better understanding of the concept that was delivered to the readers.

Limitations – Acknowledge any constraints that may have influenced the results, such as sample size, experimental conditions, biases, or methodological weaknesses. Addressing limitations demonstrates transparency and scientific rigor.

Future directions – Suggest areas for further research based on the study's findings and limitations. Proposing new questions, alternative methodologies, or expanded studies can help guide future investigations.

8.10 Conclusions

The conclusion in a manuscript is a brief, final section that summarizes the key findings, their significance, and the broader implications of the research. It reinforces the main takeaways without introducing new data or interpretations. The conclusion often highlights how the study contributes to existing knowledge, addresses the research objectives, and suggests practical applications or future research directions. A well-written conclusion leaves a lasting impression, emphasizing the study's relevance and potential impact. Some important points to be considered for drawing the conclusion

1. **Summarize key findings** – In the conclusion, it is essential to summarize key findings by briefly restating the main results and

their relevance to the research objectives. This ensures that readers clearly understand the study's core contributions, rather than repeating every result in detail, focuses on the most significant outcomes that directly answer the research questions or hypotheses. Highlight any patterns, trends, or critical insights observed in the study and explain how they align with or differ from initial expectations.

2. **Avoid repetition** – This means integrating the most important results and explaining their overall impact concisely, without repeating detailed statistical data or lengthy explanations. The conclusion should provide a big-picture perspective, showing how the findings contribute to the research field and addressing the study's objectives in a meaningful way. By summarizing the essence of the research rather than duplicating previous sections, the conclusion remains impactful, engaging, and free from redundancy.

3. **Highlight the significance** – This clearly explains how the findings add value, whether by filling research gaps, supporting or challenging existing theories, or providing new insights. If applicable, discuss how the results can be applied in real-world contexts, such as policy-making, industry practices, or technological advancements. By showcasing the study's broader impact, the conclusion reinforces its importance and relevance, leaving a strong impression on the reader.

4. **Acknowledge limitations** – It is important to acknowledge limitations to provide a balanced perspective on the study's findings. Limitations may include factors such as sample size constraints, methodological weaknesses, potential biases, or external influences that could affect the interpretation of results.

5. **Maintain clarity and conciseness** – A well-written conclusion should maintain clarity and conciseness, ensuring that the key takeaways are communicated effectively without unnecessary detail. This section should be focused on summarizing the study's main findings and their significance without introducing new data, interpretations, or lengthy explanations.

6. **End with a strong closing statement** – Ending the conclusion with a strong closing statement reinforces the study's importance and leaves a lasting impression on the reader. This final thought should concisely highlight the broader impact of the research, whether in advancing knowledge, influencing policy, or guiding future studies. A compelling closing statement can also emphasize the practical applications of the findings or suggest how they contribute to solving real-world problems.

Theoretical Questions and Problems

1. What is the role of scientific writing in knowledge dissemination?
2. How does scientific writing contribute to reproducibility in research?
3. Why is scientific writing important for establishing professional credibility?
4. In what ways does scientific writing facilitate collaboration?
5. Explain how scientific writing impacts public understanding and policy making.
6. What is meant by scientific writing providing a permanent record?
7. Describe the role of scientific writing in education and training.
8. Why is well-articulated writing important for innovation and securing funding?
9. Why is it important to define your research goals before selecting a journal?
10. How can evaluating the scope of a journal help in proper journal selection?
11. What are the indicators of a journal's reputation?
12. Compare open-access and subscription-based journals in the context of research visibility.
13. Why is checking for peer-reviewed status important when choosing a journal?
14. What are journal metrics, and how do they influence journal selection?
15. What does the indexing and archiving of a journal indicate about its quality?
16. What are predatory journals, and how can they be avoided?
17. How can colleagues and supervisors assist in journal selection?
18. List the major components of a scientific manuscript.
19. What are the essential features of a good abstract?
20. How should results be reported to maintain clarity and objectivity?
21. What are the key elements to be included in the discussion section of a manuscript?

Index

Pages in *italics* refer to figures.

For Product Safety Concerns and Information please contact our EU
representative GPSR@taylorandfrancis.com
Taylor & Francis Verlag GmbH, Kaufingerstraße 24, 80331 München, Germany

www.ingramcontent.com/pod-product-compliance
Lightning Source LLC
Chambersburg PA
CBHW070729220326
41598CB00024BA/3357